W9-AAE-676

THE WILL OF ALFRED NOBEL

The Nobel Foundation was established under the terms of the will of ALFRED BERNHARD NOBEL, Ph.D.h.c., dated Paris, November 27, 1895, which in its relevant parts runs as follows:

"The whole of my remaining realizable estate shall be dealt with in the following way: the capital, invested in safe securities by my executors, shall constitute a fund, the interest on which shall be annually distributed in the form of prizes to those who, during the preceding year, shall have conferred the greatest benefit on mankind. The said interest shall be divided into five equal parts, which shall be apportioned as follows: one part to the person who shall have made the most important discovery or invention within the field of physics; one part to the person who shall have made the most important chemical discovery or improvement; one part to the person who shall have made the most important discovery within the domain of physiology or medicine; one part to the person who shall have produced in the field of literature the most outstanding work in an ideal direction, and one part to the person who shall have done the most or the best work for fraternity between nations, for the abolition or reduction of standing armies and for the holding and promotion of peace congresses. The prizes for physics and chemistry shall be awarded by the Swedish Academy of Sciences; that for physiology or medical works by the Carolinska Institute in Stockholm; that for literature by the Academy in Stockholm, and that for champions of peace by a committee of five persons to be elected by the Norwegian Storting. It is my express wish that in awarding the prizes no consideration whatever shall be given to the nationality of the candidates, but that the most worthy shall receive the prize, whether he be Scandinavian or not."

The Nobel Prize

The First 100 Years

edited by

Agneta Wallin Levinovitz
Nils Ringertz

The Nobel Foundation
Stockholm, Sweden

ICP Imperial College Press

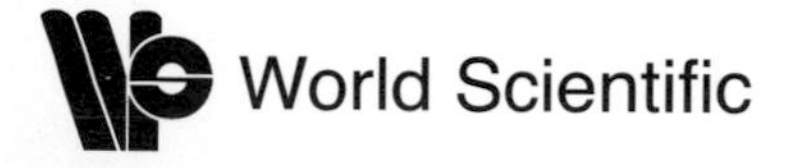

Published by

Imperial College Press
57 Shelton Street
Covent Garden
London WC2H 9HE

and

World Scientific Publishing Co. Pte. Ltd.
P O Box 128, Farrer Road, Singapore 912805
USA office: Suite 1B, 1060 Main Street, River Edge, NJ 07661
UK office: 57 Shelton Street, Covent Garden, London WC2H 9HE

British Library Cataloguing-in-Publication Data
A catalogue record for this book is available from the British Library.

THE NOBEL PRIZE: The First 100 Years

ISBN 981-02-4664-1
ISBN 981-02-4665-X (pbk)

Printed in Singapore.

Preface

In the year 2001 the Nobel Foundation celebrates the Centennial of the first Nobel Prizes. Among the events for 2001 is the opening of a Centennial Exhibition in Stockholm. An identical exhibition will open in Oslo in the fall of 2001 and then tour different cities around the world. The Nobel web site has been upgraded to Nobel e-Museum (NeM) — a virtual museum of science and culture which can be found on the Internet at www.nobelprize.org.

As part of the Centennial celebrations, the NeM is publishing a series of reviews covering the work of Nobel Laureates in Physics, Chemistry, Physiology or Medicine, Literature and Peace as well as Winners of the Bank of Sweden Prize in Economic Sciences in Memory of Alfred Nobel (awarded since 1969). Electronic publication of this series began in 1999 and has now been completed. In view of the great interest in these reviews, and to make the information available also to those who prefer to read from printed pages, a collaboration has been established with Imperial College Press and World Scientific to publish updated versions of these reviews in the form of this Centennial Volume.

We wish to thank all the contributing authors and Gudrun Franzén, administrator of the Nobel e-Museum, for her advice and help at all stages of preparing the manuscript for this volume. Thanks are also due to Dr Ola Törnkvist, Imperial College Press, London and Ms Kim Tan, World Scientific, Singapore for the copy-editing and efficient production of this volume.

Agneta Wallin Levinovitz Nils Ringertz

Contents

The Nobel Prize

The First 100 Years

Introduction

*Michael Sohlman**

The celebration of the Centennial of the Nobel Prizes in 2001 brings with it i.a. a perspective on the development of human civilization over the past hundred years. The disciplines covered by the Nobel Prizes — Physics, Chemistry, Physiology or Medicine, Literature and Peace, as well as the Bank of Sweden Prize in Economic Sciences in Memory of Alfred Nobel (from 1969) — deal with many, if not all, major aspects of the conditions of life on earth. And even if the Prizes have obviously not been able to capture all the most important contributions to the progress of Mankind, they constitute important markers of the major trends in their respective area. The articles included in this volume have the ambition to convey these major trends and developments.

When gauging the meaning and development of the Nobel Prizes, a natural starting point is to ask what Nobel himself intended with the Prizes. His intentions and the criteria he envisaged for the five disciplines are given in his last will, but they remain very broad, and as is clear from the following articles, have necessitated interpretation over the years, an interpretation which remains constantly on the agenda of the Prize-Awarding Institutions.

It is also clear that some of his intentions — that the award would go to "those who, during the preceding year, shall have conferred the greatest benefit to mankind" were impossible to fulfil already from the start: what economists call information-, recognition- and decision-lags were and are still too long.

*Executive Director of the Nobel Foundation.

We also have reason to think that Nobel — at least in the scientific disciplines — had in mind that young, talented inventors should be given a safe financial basis for their work and thereby be spared the constant trouble in finding financiers at the start of their career, as was the case for Alfred Nobel himself. Here the history shows that the Prizes have rather concentrated on the importance of the discoveries, inventions, literary works and pacifist ventures, than on encouragement of young talents. And in our days governments are financing research grants and scholarships, and a rapidly expanding venture capital market provides young start-up entrepreneurs with the needed capital.

Looking back with the perspective of a century, the question arises: What explains the present-day position of the Nobel Prizes? One answer is that the Nobel Prize when it was founded in Alfred Nobel's will, was the first truly international Prize. A number of important Prizes had been awarded in different countries on a national basis long before the Nobel Prizes. But there was no Prize with the same global and internationalist scope and mission. The Will says explicitly that "no consideration whatever shall be given to the nationality of the candidates." This was an important humanistic signal at a time when nationalism and chauvinism was on the rise. Indeed, the system of values underlying the provisions of the last will of Alfred Nobel mirrors his philosophical outlook with its combination of the ideals of the Enlightenment and strong optimism about the rapid progress of mankind. From the correspondence between Alfred Nobel and Bertha von Suttner, it is moving to learn that they thought that the eradication of war, as a kind of human behavior, would be a matter of 20–30 years, i.e. approximately around 1914.

The main reason for the standing of the Prize today is, however, the importance of the names on the list of Laureates and their contributions to human development. And it has been the difficult task of the Prize-Awarding Institutions over the years, to interpret the last will of Alfred Nobel in the light of a constantly changing world. In this work they are assisted in a decisive way by colleagues and experts from all over the world, who participate in the award process, either as nominators and/or as contributors to the evaluation of the different candidates. This wide network of contacts gives the Prizes the character of recognition by peers in the respective field.

On the threshold of the next century of Nobel Prizes, the Prize-Awarders face the daunting task to combine the criteria and formal limits of Nobel's last will with the ever-changing reality of science, literature, and striving for peace.

As a sign of our times, this volume consists of overview articles first written for the official web site of the Nobel Foundation — www.nobel.se. By being printed in this volume the unstoppable progress of Mankind from 'Gutenberg to Gates' has been temporarily halted for the benefit of readers who still enjoy holding a book in their hands.

Life and Philosophy of Alfred Nobel[*]

Tore Frängsmyr[**]

Probably no Swede is as well-known throughout the world as Alfred Nobel — not our medieval saints, nor even our contemporary sports heroes. At the same time, we must admit that his renown is more indirect than direct. This means that while the Nobel Prize is extremely well-known all over the world, the person behind it remains relatively unknown.

Admittedly, quite a lot has been written about Alfred Nobel, but a large part of this literature consists of clichés. It is often a question of sentimental depictions of a lonely millionaire who — despite his wealth — was unhappy or at least deeply melancholic, emotionally attached to his mother, and with a few heart-rending love stories behind him. This is not altogether a false picture. Alfred Nobel was lonely and he was clearly unlucky in love, but such accounts are not so instructive. Romantic tales constitute a special genre, to which I shall not attempt to contribute. Instead, I will focus on the scientific and technical fields.

First, however, I would like to recount some important facts about Alfred Nobel's life. He was born in Stockholm in 1833 into a family of engineers. His family was descended from none other than Olof Rudbeck, the best-known technical genius of Sweden's 17th century era as a Great Power in Northern Europe. Having gone through a recent bankruptcy, when Alfred was five years old his father Immanuel Nobel moved to St. Petersburg, where he started a mechanical workshop for the manufacture

[*]Memorial address at the Royal Swedish Academy of Sciences, March 26, 1996.
[**]Director of the Center for History of Science at the Royal Swedish Academy of Sciences.

of land mines. In 1842, when Alfred was nine years old, the rest of the family also moved to St. Petersburg. By then his father's fortunes had improved, enabling the family to live in high bourgeois style. At the time, St. Petersburg was a world metropolis, alive with scientific, social, and cultural life. Immanuel Nobel's sons did not attend school, but were instead educated at home by outstanding teachers at the level of university professor. The instruction they provided focused on both the humanities and the natural sciences. Aside from Swedish, Alfred and his brothers were taught Russian, French, English and German, as well as literature and philosophy. In the natural sciences, they were guided by two professors of chemistry who taught them mathematics, physics and chemistry. Considering the specialty of his teachers, it was perhaps no coincidence that Alfred took a liking to chemistry. He learned to conduct chemical experiments, an activity that seemed to fascinate him from the very beginning. Alfred spent his most important formative years in the Russian capital. With his five languages, which he seemed to have mastered well, he laid the foundation for the cosmopolitan nature that would later become so prominent in his life.

During the years 1850–1852, Alfred was allowed a few study-oriented stays abroad. He spent one year in Paris with the famous chemist Jules Pelouze, a professor at the Collège de France who had just opened a private training laboratory. Pelouze, who incidentally had been a good friend of the Swedish chemist Berzelius, had also taught Nikolai Zinin, one of Alfred Nobel's private teachers. During that year, Alfred completed his training as a chemist. But somewhere around the same time was the inception of what would become the greatest inventions of his life. For it was then, if not earlier, that he must have heard about the remarkable explosive called nitroglycerine. Strangely enough, this has not been pointed out by many scholars, who have dated the crucial moment 10 years later.

Here is the background. In 1847, in Turin, Ascanio Sobrero — an Italian student of Pelouze — had discovered a new explosive that he initially called pyroglycerine (later known as nitroglycerine). However, Sobrero, both in letters to Pelouze and in a subsequent journal article, issued a warning about the new compound, not only because it had incredible explosive power, but also because it was impossible to handle. Sobrero's discovery did not come as a bolt from the blue. As early as the 1830s, Pelouze himself and others had conducted important preliminary work by making guncotton. Since Alfred was extremely interested in explosives — it was of course a family interest — and since Pelouze had both first-hand knowledge of how explosives were manufactured and was familiar with Sobrero's discovery, Alfred must have learned about nitroglycerine at that time. However, any excitement he might have felt was immediately dampened by the difficulties of both manufacturing and handling the new compound.

The end of the Crimean War (1856) spelled disaster for Immanuel Nobel's factory, which had lived off the manufacture of war materiel. The factory went bankrupt, and Alfred's parents and their youngest son Emil moved back to Sweden. The three older sons stayed in St. Petersburg to put the family affairs in order and restructure the company. Faced with this situation, Alfred and his brothers discussed various conceivable projects with their former teachers. That was when Nikolai Zinin reminded them of the potential of nitroglycerine. Professor Zinin is said to have demonstrated the power of nitroglycerine by pouring a few drops of the fluid on an anvil, striking it with a hammer, and producing a loud bang. But only the liquid that came into contact with the hammer exploded. The rest of the liquid was not affected. The problem, as Sobrero had already realized, was two-fold. First, it was difficult to manufacture the compound, because at excessive temperatures the whole batch exploded. Second, once manufactured, the liquid was equally difficult to explode in a controlled fashion.

During the years around 1860, Alfred conducted repeated experiments involving great risks. First, he succeeded in manufacturing sufficient quantities of nitroglycerine without any mishaps. Then, he mixed nitroglycerine with black gunpowder and ignited the mixture with an ordinary fuse. After several successful explosions outside St. Petersburg on the frozen Neva River, Alfred traveled back to Stockholm. There, his father had begun similar experiments (though with less success) after reading about Alfred's tests in his letters. Immanuel Nobel even insisted that the new mixture was his own idea, but he backed off from this assertion after a sharp letter from Alfred that set matters straight in no uncertain terms. Instead, he even helped Alfred apply for a patent in his own name. In October 1863, Alfred Nobel was granted a patent for the explosive that he aptly called 'blasting oil'.

With his first patent, Alfred had also reached his first milestone. Although he was only 30 years old, this was the start of an exciting adventure that would unfold with great speed. During the following spring and summer, Alfred continued his experiments. He soon obtained a new patent related to the manufacture of nitroglycerine (using a simplified method) as well as the use of a detonator, or what was called an 'initial igniter', in other words a hollow wooden plug filled with black gunpowder (later called a 'blasting cap'). The determination and self-confidence that would later become more pronounced features of Alfred's personality were already apparent. He wrote: "I am the first to have brought these subjects from the area of science to that of industry," and he successfully arranged a large loan from a French bank.

Around the same time, another personality trait began to assert itself — the inventor also became an entrepreneur. Alfred dealt with failures in the same resolute manner as he did successes. In September 1864, a major explosion at the Nobel factory in Stockholm claimed the lives of Alfred's brother Emil and four other people. Just one month later, Alfred — resolutely and without sentimentality — founded his first joint stock company. Despite the accident or perhaps because of it, since no one could now doubt the explosive power of the new compound, orders began rolling in. The Swedish State Railways ordered blasting oil for use in building the Söder Tunnel in Stockholm. A year later, in 1865, Alfred improved his blasting cap (now made of metal rather than wood) which in principle is still of the same type used today. He then left for Germany, set up a company there and bought land outside Hamburg where he built a factory. In the summer of 1866, Alfred Nobel traveled to America. There he struggled against political bureaucracy, popular fear of accidents caused by explosives and, not least, dishonest business associates. In the end, he received patents, formed companies and built factories there.

Despite slow communications, everything now happened very quickly. Events literally assumed explosive force. While Alfred was in America, his factory in Germany exploded. When he returned to Germany in August, he had to supervise the clean-up of the debris and plan a new building. At the same time, he continued to brood over the safety problems of nitroglycerine and he conducted new experiments. He realized that nitroglycerine had to be absorbed by some kind of porous material, forming a mixture that would be easier to handle. On the German moorlands very close to where he was staying, he found a type of porous, absorbent sand or diatomaceous earth known in German as Kieselguhr. When nitroglycerine was absorbed by Kieselguhr, it formed a paste that was easy to knead and shape. This paste could be shaped into rods that were easily inserted into drilling holes. It could also be transported and subjected to jolts without triggering explosions. It could even be ignited without anything happening. Only a blasting cap would cause the paste to explode. The disadvantage of this new substance was its somewhat reduced explosive force — the Kieselguhr did not participate as an active substance in the explosion. But this was the price one had to pay. In short, that was how Alfred Nobel invented dynamite. Incidentally, Alfred himself coined the word dynamite from the Greek dynamis, meaning power. One of his German colleagues had proposed the term 'blasting putty' because it had the same consistency as putty. But Alfred thought this sounded like something meant to be used for blasting window panes, which was certainly not his intention. In 1867, he was granted patents for dynamite in various countries, notably Britain, Sweden and the United States. Production was now set to begin on a large scale, and demand grew rapidly. It was an era of large infrastructure projects like railways, ports, bridges, roads, mines and tunnels, where blasting was necessary. For example, dynamite was of vital importance in the construction of the St. Gotthard tunnel through the Swiss Alps in the 1870s.

In 1868, the year after the first patent for dynamite, Alfred Nobel and his father were awarded the Letterstedt Prize by the Royal Swedish Academy of Sciences. This prize, which Alfred valued highly, was awarded for "important discoveries of practical value to humanity." We can hear an echo of this wording in Nobel's will, where he stated the criteria for awarding his own prizes.

He had taken the decisive steps that led to honor and fame. Let us pause a moment at the year 1873, when Alfred Nobel was 40 years old. All these events had taken place during the preceding 10 years. At age 30, Alfred had received his first patent. Now, by age 40, he had already made his greatest discoveries, he had built up a worldwide industrial empire, he had become wealthy, and he had bought a large house in the center of Paris. The foundation was in place. He later made new discoveries — primarily blasting gelatin and ballistite — and his industrial enterprises, as well as his fortune, grew. His distinguishing quality was his versatility. He was an inventor, an industrialist and an administrator. He had to safeguard his patent rights, develop products, establish new companies, and conduct business in five languages with the rest of the world — without the help of a secretary and before the telephone and fax made people's lives easier. He frequently traveled by train or boat, since this was before the advent of the airplane. His factories exploded, he had to withstand negative publicity campaigns, and he unmasked deceitful business partners. He had to deal with all of this himself. In addition, he seldom felt well — he viewed himself as sickly and frail, often complaining of migraines, rheumatism and an unsettled stomach. His life was hectic and stressful. In letters he wrote from Paris, he complained of being constantly hounded by people, which he described in his own words as "pure torture." People are crazy, he wrote — they rushed in and out of his office, everyone wanted to see him, and his presence was required everywhere. But despite everything, he managed to cope. In the role of the entrepreneur, he was unbeatable.

I would like to touch upon another level of Alfred Nobel's personality, that of the humanist and philosopher. We know that he had literary interests and ambitions. He was an avid reader of fiction and wrote his own dramatic works and poems. In addition, he was attracted to philosophical issues. He read certain philosophical works with such interest that he underlined important passages. Among the papers that he left behind is a black notebook on philosophy that his biographers have not taken an interest in. Although not constituting profound original thoughts, these penciled notes reflect his serious interest in philosophical questions. Nobel went through philosophy from antiquity to modern times, pointing out what he perceived to be vital issues. He made his own comments, which in a morose way showed his detachment from the subject. He commented on Plato, Aristotle and Democritus, but also on Newton and Voltaire as well as contemporary biologists such as Darwin and Haeckel. Nobel noted, for example, that it was unclear what caused people to form a conception of a God: "Aristotle attributes it to fear, Voltaire to the desire of the more clever to deceive the stupid." He spoke with respect of the philosophical doubts of Descartes and Spinoza, adding that doubt must surely be the starting point for all philosophical thinking. Theories of knowledge were of special interest to Nobel. Consequently, he returned several times to Locke's thesis that all knowledge arises from sensory impressions, declaring that the "brain is a very unreliable recorder of impressions."

This led him to reflect further on the methodology of science and to develop a line of reasoning that, aside from being inspired by Locke's thesis, also seemed to have been influenced by Alexander von Humboldt's theory of knowledge. Nobel wrote that all science is built on observations of similarities and differences. He continued:

> "A chemical analysis is of course nothing other than this, and even mathematics has no other foundation. History is a picture of past similarities and differences; geography shows the differences in the earth's surface; geology, similarities and differences in the earth's formation, from which we deduce the course of its transformations. Astronomy is the study of similarities and differences between celestial bodies; physics, a study of similarities and differences that arise from the attraction and motive functions of matter. The only exception to this rule is religious doctrine, but even this rests on the similar gullibility of most people. Even metaphysics — if it is not too insane — must find support for its hypotheses in some kind of analogy. One can state, without exaggeration, that the observation of

and the search for similarities and differences are the basis of all human knowledge."

Nobel could have completed this train of thought with Humboldt's words that "from observation one goes on to experimentation based on analogies and inductions of empirical laws." Nobel did not espouse any grand theory of knowledge, but rather an empirical method. Alfred Nobel himself seemed to think that he had accomplished quite a lot by applying this method in his work.

Alfred Nobel also viewed himself with detachment, or shall we say, philosophical skepticism. He often described himself as a loner, hermit, melancholic or misanthrope. He once wrote: "I am a misanthrope and yet utterly benevolent, have more than one screw loose yet am a super-idealist who digests philosophy more efficiently than food." Even from this description, it is clear that this misanthrope was also a philanthropist, or what Nobel called a super-idealist. It was the idealist in him that drove Nobel to bequeath his fortune to those who had benefited humanity through science, literature and efforts to promote peace.

For Alfred Nobel, the idea of giving away his fortune was no passing fancy. He had thought about it for a long time and had even re-written his will on various occasions in order to weigh different wordings against each other. Efforts to promote peace were close to his heart, largely inspired by his contacts with Bertha von Suttner (herself a Nobel Peace Prize winner in 1905). He derived intellectual pleasure from literature, while science built the foundation for his own activities as a technological researcher and inventor. On November 27, 1895, Nobel signed his final will and testament at the Swedish–Norwegian Club in Paris.

Alfred Nobel had many different homes during the final decades of his life. In 1891, he had left Paris to live in San Remo, Italy, after controversies with the French authorities. Four years later, he purchased the Bofors ironworks and armaments factory in Sweden and established his Swedish home at nearby Björkborn Manor. He equipped all his residences with laboratories where he could continue his experiments. He was apparently homesick for Sweden but complained of the Swedish winter weather. His health began to falter. He visited doctors and health resorts more frequently, but never had time to heed their most important advice — "to rest and nurse my health," as he put it himself. On December 10, 1896, Alfred Nobel passed away at his home in San Remo.

Nobel's will was hardly longer than one ordinary page. After listing bequests to relatives and other people close to him, Nobel declared that his entire remaining estate should be used to endow "prizes to those who, during the preceding year, shall have conferred the greatest benefit on mankind." His will attracted attention throughout the world. It was unusual at that time to donate large sums of money for scientific and charitable purposes. Many people also criticized the international character of the prizes, saying they should be restricted to Swedes. This would not have suited the cosmopolitan Alfred Nobel. Some of his relatives contested the will. Complicated legal and administrative matters also had to be sorted out. All this took time, but eventually it was all settled.

In 1901, the first Nobel Prizes were awarded. The donor himself could hardly have dreamed of the impact that his benevolence would have in the future.

The Nobel Foundation: A Century of Growth and Change

*Birgitta Lemmel**

On June 29, 2000, the Nobel Foundation celebrated its 100th anniversary. The Foundation and especially the Nobel Prizes — which were first awarded in 1901 — are closely linked to the history of modern science, the arts and political development throughout the 20th century.

1. Background and Establishment of the Nobel Foundation

Alfred Nobel died on December 10, 1896. The provisions of his will and their unusual purpose, as well as their partly incomplete form, attracted great attention and soon led to skepticism and criticism, also aimed at the testator due to his international spirit. Only after several years of negotiations and often rather bitter conflicts, and after various obstacles had been circumvented or overcome, could the fundamental concepts presented in the will assume solid form with the establishment of the Nobel Foundation. On April 26, 1897, the Storting (Norwegian Parliament) approved the will. In 1898 the other prize-awarding bodies followed suit, approving the will after mediation: Karolinska Institutet on June 7, the Swedish Academy on June 9 and the Royal Swedish Academy of Sciences on June 11.

The will was now settled. The task of achieving unity among all the affected parties on how to put its provisions into practice remained. The final version of the Statutes of the Nobel Foundation contained clarifications of the wording of the will and a provision that prizes not considered

*Head of Information of the Nobel Foundation in 1986–1996.

possible to award could be allocated to funds that would otherwise promote the intentions of the testator. The Statutes provided for the establishment of Nobel Committees to perform prize adjudication work and Nobel Institutes to support this work, as well as the appointment of a Board of Directors in charge of the Foundation's financial and administrative management.

On June 29, 1900, the Statutes of the newly created legatee, the Nobel Foundation, and special regulations for the Swedish Prize-Awarding Institutions were promulgated by the King in Council (Oscar II). The same year as the political union between Sweden and Norway was dissolved in 1905, special regulations were adopted on April 10, 1905, by the Nobel Committee of the Storting (known since January 1, 1977 as the Norwegian Nobel Committee), the awarder of the Nobel Peace Prize.

2. Premises

To create a worthy framework around the prizes, the Board decided at an early stage that it would erect its own building in Stockholm, which would include a hall for the Prize Award Ceremony and Banquet as well as its own administrative offices. Ferdinand Boberg was selected as the architect. He presented an ambitious proposal for a Nobel Palace, which generated extensive publicity but also led to doubts and questions. World War I broke out before any decision could be made. The proposal was 'put on ice' and by the time the matter was revived after the war, Ivar Tengbom was busily designing what later became the Stockholm Concert Hall. Meanwhile the Stockholm City Hall was being built under the supervision of Ragnar Östberg. Boberg, Tengbom and Östberg were probably the most respected architects in Sweden at that time. Because it would have access to both these buildings for its events, the Nobel Foundation now only needed space for its administrative offices. On December 19, 1918, a building at Sturegatan 14 was bought for this purpose. After years of renovation there, the Foundation finally left its cramped premises at Norrlandsgatan 6 in 1926 and moved to Sturegatan 14, where the Foundation has been housed ever since.

3. Objectives of the Foundation

The Nobel Foundation is a private institution. It is entrusted with protecting the common interests of the Prize-Awarding Institutions named in the will, as well as representing the Nobel institutions externally. This includes informational activities as well as arrangements related to the presentation of the Nobel Prizes. The Foundation is not, however, involved in the selection process and the final choice of the Laureates (as Nobel Prize winners are also called). In this work, the Prize-Awarding Institutions are not only entirely independent of all government agencies and organizations, but also of the Nobel Foundation. Their autonomy is of crucial importance to the objectivity and quality of their prize decisions. One vital task of the Foundation is to manage its assets in such a way as to safeguard the financial base of the prizes themselves and of the prize selection process.

4. Statutes and Significant Amendments during 100 Years

The Statutes, as most recently revised in 2000, assign roles to the following bodies or individuals in the Nobel Foundation's activities:

- The Board and the Executive Director (especially Paragraphs 13 and 14)
- The Prize-Awarding Institutions (especially Paragraphs 1 and 2)
- The Trustees of the Prize-Awarding Institutions (especially Paragraph 18)
- The Nobel Committees and experts (especially Paragraph 6)
- Bodies and individuals entitled to submit prize nominations (especially Paragraph 7)
- Auditors (especially Paragraph 19)

Over the past 100 years, there have been a number of changes in the relationship between the Foundation's Board of Directors and the Swedish State. Their links have gradually been severed.

According to Paragraph 14 of the first Statutes from 1901, the Foundation was to be represented by a Board with its seat in Stockholm, consisting of five Swedish men. One of these, the Chairman of the Board, was to be designated by the King in Council. The Trustees of the Prize-Awarding Institutions would appoint the others. The Board would choose an Executive Director from among its own members. An alternate (deputy) to the Chairman would be appointed by the King in Council (effective in 1974, by the Government), and two deputies for the other members would be elected by the Trustees. Since 1995 the Trustees have appointed all members and deputies of the Board. The Board chooses a Chairman, Deputy Chairman and Executive Director from among its own members.

The first Board of Directors of the Nobel Foundation was elected on September 27, 1900. On the following day, former Prime Minister Erik Gustaf Boström was appointed Chairman of the Board by the King in

Council. Effective on January 1, 1901 the Board assumed management of the Foundation's assets.

Until 1960 the Chairman was chosen from the small group of 'Gentlemen of the Realm' — prime ministers, ministers for foreign affairs and other high officials. In 1960 for the first time, a renowned scientist was chosen: Arne Tiselius, Professor of Biochemistry at Uppsala University and 1948 Nobel Laureate in Chemistry. Since then the Chairman has been chosen from among members of the Prize-Awarding Institutions. It has also become a rule that the Deputy Chairman as well as one of the members of the Board elected by the Trustees should be persons with financial expertise. In most cases, the Executive Director has had a legal and administrative background. As the Foundation's investment policy became more active from the early 1950s onward, financial experience and a knowledge of international relations have become a necessity for those holding this position.

An important landmark in the history of the Foundation occurred when it added Norwegian representation to the Board. In 1901, the Norwegians refrained from representation on the Board — being appointed by King Oscar at a time when Norway was moving toward a breakup of its union with Sweden was not considered an attractive idea — and they limited their involvement to work as trustees and auditors. In light of this, it is interesting to note that Henrik Santesson, the first Executive Director of the Foundation, also happened to be the legal counsel of the Storting in Sweden. But in 1986, Paragraph 14 of the Statutes was changed and the Board no longer had to consist of five Swedish citizens (the original Statutes had said Swedish men), but of six Swedish or Norwegian citizens. The Statutes were also changed in such a way that remuneration to the Board members and auditors of the Foundation, as well as the salary of the Executive Director, would be determined by the Foundation's Board instead of the Swedish Government.

According to Paragraph 17 of the original Statutes, the administration of the Board and the accounts of the Foundation for each calendar year were to be examined by five auditors. Each prize-awarding body would elect one of these before the end of the year and the King would designate one, who would be the chairman of the auditors. In 1955 the number of auditors was enlarged from five to six; the new auditor would be appointed by the Trustees and had to be an authorized public accountant. This was a very important change, in line with the Foundation's more active financial investment policy.

Today the Government's only role in the Nobel Foundation is to appoint one auditor, who is also to be the chairman of the Foundation's auditors.

Among other changes that have occurred in the Statutes are the following:

Until 1968, in principle more than three persons could share a Nobel Prize, but this never occurred in practice. The previous wording of Paragraph 4 was: "A prize may be equally divided between two works, each of which may be considered to merit a prize. If a work which is to be rewarded has been produced by two or more persons together, the prize shall be awarded to them jointly." In 1968 this section was changed to read that "In no case may a prize be divided between more than three persons."

In 1974, the Statutes were changed in two respects. The confidential archive material that formed the basis for the evaluation and selection of candidates for the prizes, which was previously closed to all outsiders, could now be made available for purposes of historical research if at least 50 years had elapsed since the decision in question. The other change concerned deceased persons. Previously, a person could be awarded a prize posthumously if he/she had already been nominated (before February 1 of the same year), which was true of Erik Axel Karlfeldt (Literature Prize, 1931) and Dag Hammarskjöld (Peace Prize, 1961). Effective from 1974, the prize may only go to a deceased person to whom it was already awarded (usually in October) but who had died before he/she could receive the prize on December 10 (William Vickrey, 1996 Prize in Economic Sciences in Memory of Alfred Nobel).

5. Financial Management

The main task of the Nobel Foundation is to safeguard the financial base of the Nobel Prizes and of the work connected to the selection of the Nobel Laureates.

In its role as a financial manager, the Nobel Foundation resembles an investment company. The investment policy of the Foundation is naturally of the greatest importance in preserving and increasing its funds, thereby ensuring the size of the Nobel Prizes. The provisions of Alfred Nobel's will instructed his executors to invest his remaining realizable estate, which would constitute the capital of what eventually became the Nobel Foundation, in 'safe securities'. In the original by-laws of the Board, approved by the King in Council on February 15, 1901, the expression 'safe securities' was interpreted in the spirit of that time as referring mainly to bonds or loans — Swedish as well as foreign — paying fixed interest and backed by solid underlying security (central or local government, property mortgages or the like). In those days, many bonds were sold with a so-called gold clause, stipulating that the holder was entitled to demand payment in gold. The stock market and real estate holdings were beyond the pale. Stocks in particular were regarded as an excessively risky and speculative form of financial investment.

The first 50 years of management came to be characterized by rigidity in terms of financial investments and by an increasingly onerous tax burden. Remarkably, the tax issue had not been addressed when the Nobel Foundation was established. The tax-exempt status that the executors of the will and others had assumed as self-evident was not granted. Until 1914, the tax was not excessively heavy, only 10 percent, but when a 'temporary defense tax' supplement was introduced in 1915, the Foundation's tax burden doubled. In 1922, a maximum tax assessment was imposed which exceeded the sum available for the prizes in 1923, the year when the Nobel Prize amount reached its absolute low point. For a long time, the Nobel Foundation was the largest single taxpayer in Stockholm. The question of granting tax-exempt status to the Foundation was debated back and forth in the Riksdag (Swedish Parliament) for years.

In 1946, when the Foundation was finally exempted from national income and wealth tax and local income tax, this allowed a gradual long-term increase in the size of the Foundation's main fund, the Nobel Prizes and the sums paid to the Prize-Awarding Institutions for their adjudication work. Without Swedish tax-exempt status, it would have been impossible for the Foundation to receive equivalent tax relief for its financial investments in the United States. In the event, a US Treasury ruling granted the Foundation tax-exempt status in that country effective from 1953. Tax-exempt status created greater freedom of action, enabling the Foundation to pursue an investment policy not dominated by tax considerations that characterize the actions of many investors.

However, the restrictions on the Foundation's freedom of investments continued with minor changes until 1953, although the gold clause and resulting protection against declining value had disappeared as early as World War I. Because of two world wars and the depression of the early 1930s, the prizes shrank in real terms from SEK 150,000 in 1901 (equivalent to 20 times the annual salary of a university professor) to a mere one-third of this value.

Then, in 1953, the Government approved a radical liberalization of the investment rules. The Foundation was granted more extensive freedom to manage its capital independently, as well as the opportunity to invest in stocks and real estate. Freedom of investment, coupled with tax-exemption and the financial expertise of the Board, led to a transformation from passive to active management. This can be regarded as a landmark change in the role of the Foundation's Board. During the 1960s and 1970s, the value of the Nobel Prizes multiplied in Swedish krona terms but rapid inflation meanwhile undermined their real value, leaving each prize largely unchanged. The same was true of the Foundation's capital.

During the 1980s, the Foundation experienced a change for the better. The stock market performed outstandingly and the Foundation's real estate also climbed in value. A sour note came in 1985, when Swedish real estate taxes rose sharply and profits consequently vanished. In 1987, the Board decided to transfer most of the Foundation's real estate to a separate company called Beväringen, which was then floated on the stock exchange. In the same year that Beväringen was established, the Nobel Foundation surpassed its original value in real terms (SEK 31 million in 1901 money) for the first time. The Foundation was fortunate enough to sell its entire holding in Beväringen before the real estate crash of the early 1990s.

By 1991, the Foundation had restored the Nobel Prizes to their 1901 real value. Today the nominal fund capital of the Nobel Foundation is about SEK 4 billion. Each of the five Nobel Prizes as well as the Economics Prize will, in 2001, be worth SEK 10 million (about USD 1 million). This is well above the nominal value of the entire original fund, and higher than the real value of the original prizes. Since January 1, 2000, the Nobel Foundation has also been permitted to apply the capital gains from the sale of assets toward the prize amounts. According to Alfred Nobel's will, only direct return interest and dividends could be used for the prize amounts. Capital gains from share management could not previously be used. According to the new rules, return that arises from the sale of Foundation assets may also be used for prize award events and overhead, to the extent that they are not needed to maintain a good long-term prize-awarding capacity. This change is necessary to avoid undermining the value of the Nobel Prizes. The Nobel Foundation may also decide how much of its assets may be invested in shares. In the long term, this may mean that the Foundation can now have a higher percentage of its assets invested in shares, leading to higher overall return and thus larger Nobel Prizes.

6. The Sveriges Riksbank (Bank of Sweden) Prize in Economic Sciences in Memory of Alfred Nobel

On the occasion of its 300th anniversary in 1968, the Bank of Sweden (Sveriges Riksbank) made a large donation to the Nobel Foundation. A Prize in Economic Sciences in Memory of Alfred Nobel has been awarded since 1969. The Royal Swedish Academy of Sciences is entrusted with the role of Prize-Awarding Institution, in accordance with Nobel Prize rules. The Board of the Nobel Foundation has subsequently decided that it will allow no further new prizes.

7. Nobel Symposia

An important addition to the activities of the Nobel Foundation is its Symposium program, which was initiated in 1965 and has achieved a high international standing. Since then 120 Nobel Symposia, dealing with topics at the frontiers of science and culture and related to the Prize categories, have taken place. Since 1982 the Nobel Symposia have been financed by the Foundation's Symposium Fund, created in 1982 through an initial donation from the Bank of Sweden Tercentenary Foundation and the Knut and Alice Wallenberg Foundation, as well as through grants and royalties received by the Nobel Foundation as part of its informational activities.

8. Donations and Prizes

Around the world, new international scientific and cultural prizes have been established, directly inspired by the Nobel Prize. For example, the Japan Prize and Kyoto Prize — both financially in a class with the Nobel Prize — were established in 1985 and their statutes directly refer to the Nobel Prizes as a model and source of inspiration. Donations from these and many other sources have reached the Foundation over the years. Some of these donations are presented below.

In 1962 the Balzan Foundation, based in Switzerland and Italy, gave its first prize of one million Swiss francs to the Nobel Foundation for having awarded its Nobel Prizes for 60 years in an exemplary way, thereby celebrating "l'oeuvre admirable accomplie dans 60 années de travail."

In 1972, Georg von Békésy, 1961 Nobel Laureate in Physiology or Medicine, donated his exquisite collection of art objects to the Nobel Foundation — some 150 objects from four continents (not Australia).

Also in 1972 the Foundation received a donation from the Italian marquis Luigi de Beaumont Bonelli, who bequeathed his two wine-growing estates outside Taranto, southern Italy, to the Nobel Foundation. The properties were worth SEK 4.5 million. Their sale made possible the establishment of an annual Beaumont–Bonelli fellowship to a promising young Italian medical researcher.

As to the two Japanese prizes mentioned earlier, on April 20, 1985, the Science and Technology Foundation of Japan established the Japan Prize. At the first award ceremony, a special prize of JPY 50 million was awarded to the Nobel Foundation "in recognition of the role the Nobel Foundation has played since 1901 in promoting science and international understanding." On November 10, 1985, the Inamori Foundation in Kyoto awarded its first Kyoto Prize of JPY 45 million to the Nobel Foundation "with the aim of promoting science, technology and the arts in the spirit of the Nobel Prize."

9. Nobel Festivities

The Nobel Foundation is an 'investment company' with rather unusual facets. Every year this investment company moves into show business by organizing the Nobel Festivities and numerous related arrangements that take place in December. The Nobel Foundation is responsible for organizing the Nobel Festivities in Stockholm, while in Norway the Norwegian Nobel Committee is in charge of the corresponding arrangements. On December 10, 1901, the Nobel Prizes were awarded for the first time in Stockholm and in Christiania (now Oslo) respectively.

9.1. *Stockholm*

The Prize Award Ceremony in Stockholm took place at the Old Royal Academy of Music during the years 1901–1925. Parenthetically, it is worth mentioning that during the first years the names of the Nobel Laureates were not made public until the Award Ceremony itself.

Since 1926, the Prize Award Ceremony has taken place at the Stockholm Concert Hall with few exceptions, last time in 1991 at the Stockholm Globe Arena, when the 90th anniversary of the first Nobel Prizes was the focus of the celebrations.

Until the early 1930s, the Nobel Banquet took place at the Hall of Mirrors in the Grand Hotel, Stockholm. In its very first years, 1901 and

1902, the Banquet was an exclusive party for men only. Once the Stockholm City Hall had been built, in 1930 a decision was made to hold the Banquet in its fantastic Golden Hall this year and in the future. Over time, the character of the Banquets changed and interest in participating became greater and greater. Starting in 1974, due to the need for more space the Nobel Banquet was moved from the Golden Hall to the larger Blue Hall of the City Hall, which today accommodates some 1,300 guests. The Blue Hall had only been used for the Banquet once before, in 1950, when the Nobel Foundation celebrated its 50th anniversary.

There are always exceptions to the rules. In 1907, there were no festivities in Stockholm because the Royal Court was in mourning. King Oscar II had just died. The Laureates were awarded their prizes at a ceremony at the auditorium of the Royal Swedish Academy of Sciences. During 1914–1918 the Nobel Festivities were called off in Sweden and in Norway, except for a ceremony in 1917 at the Norwegian Nobel Institute in the presence of King Haakon to announce that the International Red Cross had been awarded the Peace Prize.

The first Nobel Prizes after World War I — the 1919 prizes — were awarded in June the next year in order to give the Festivities an atmosphere of early Swedish summer with sunshine, light and greenery instead of dark December with cold and wet snow. The events took place on June 2, 1920 but it was not a success. No members of the Royal Family were present because of the death of Crown Princess Margaretha. The weather was gray, rainy and cold. As a result of disappointment the Nobel Festivities of 1920 reverted to earlier tradition and were held on December 10.

In 1924 the Nobel Festivities were canceled in Stockholm. Neither of the two Laureates could be present: the Laureate in Physiology or Medicine was traveling and the Literature Laureate was unwell. The Prizes in Physics and Chemistry were reserved that year.

During the period 1939–1943, the Nobel Festivities were called off. In 1939 only the Laureate in Literature, Frans Eemil Sillanpää from Finland, received his Prize in Stockholm at a small ceremony. During 1940–1942 no Physics, Chemistry or Medicine Prizes were awarded, during 1940–1943 no Literature Prizes, and during 1939–1943 no Peace Prizes.

In 1944 there were no festivities in Stockholm, but a luncheon was held at the Waldorf–Astoria Hotel in New York organized by the American Scandinavian Foundation. Some 1943 and 1944 Laureates received their Prizes from the Swedish Minister in Washington, two Physics Laureates — Otto Stern (1943) and Isidor Isaac Rabi (1944) — and four Laureates in Physiology or Medicine — Henrik Dam and Edward Doisy (1943), and

Joseph Erlanger and Herbert S. Gasser (1944). A speech by Sweden's Crown Prince Gustaf Adolf was broadcast on American radio the same day. The 1943 Laureate in Chemistry, George de Hevesy, received his Prize in Sweden without any ceremonies and the 1944 Literature Laureate, Johannes V. Jensen from Denmark, received his Prize in Stockholm in 1945.

Just before and during the war, Adolf Hitler forbade Laureates from Germany — Richard Kuhn (Chemistry, 1938), Adolf Friedrich Johan Butenandt (Chemistry, 1939), Gerhard Domagk (Physiology or Medicine, 1939) and Otto Hahn (Chemistry, 1944) — from accepting their Prizes at that time. However, they received their insignia on later occasions.

In 1956, due to the crisis in Hungary, a smaller, more private dinner at the Swedish Academy replaced the glittering banquet in the City Hall, although the Prize Award Ceremony took place as usual at the Concert Hall.

9.2. *Christiania/Oslo*

In Norway, during the years 1901–1904 the decision on the Peace Prize was announced at a meeting of the Storting on December 10, after which the recipients were informed in writing. On December 10, 1905, the Nobel Institute's new building at Drammensveien 19 was inaugurated in the presence of the Norwegian Royal Couple, and it was announced that Bertha von Suttner had received the 1905 Peace Prize. The Laureate herself was not present. During 1905–1946 the Prize Award Ceremonies were held at the Nobel Institute building, during 1947–1989 in the auditorium of the University of Oslo and since 1990 at the Oslo City Hall. The King of Norway is present, but it is the Chairman of the Nobel Committee who hands over the Prize to the Laureate or Laureates. The Nobel Banquet in Norway is a dignified formal occasion, but much less pretentious than the Banquet in Stockholm. It takes place at the Grand Hotel in Oslo, with approximately 250 guests.

10. The Norwegian Nobel Committee and the Nobel Foundation during World War II

In 1940, three members of the Storting's Nobel Committee were in exile due to the occupation of Norway by Nazi Germany, which lasted until 1945. The remaining members and deputies kept the work of the Committee going. Because the Storting could not elect new Committee members, the Nobel Foundation asked existing members to continue in their posts.

In January 1944, pro-Nazi Prime Minister Vidkun Quisling and his administration wanted to take over the functions of the Nobel Committee in Norway and seize control of the Nobel Institute's building on Drammensveien. After consultations with the Swedish Foreign Ministry and the Director of the Nobel Institute, the Nobel Foundation declared that the Nobel Institute was Swedish property. Those Committee members who had remained in Norway stated in writing that under the prevailing circumstances, they could not continue their work. Sweden's consul general in Oslo, who had already moved into an office on the Nobel Institute's premises, took over the management of the building and the functions of the Nobel Institute. In 1944–1945 the Nobel Foundation together with the members of the Nobel Committee in exile ensured that nominations were submitted for the 1945 Peace Prize.

11. A New Century

After a hundred years of existence, the Nobel Prizes — as well as the centenarian Nobel Foundation — have become solid institutions, based on a great tradition since their beginning. The original criticisms aimed at the whole idea of the Nobel Prizes have faded into oblivion. Both in Sweden and in Norway, the awarding of the prizes is regarded as an event of national importance. The Nobel Foundation has now entered a new century, with museum and exhibition projects underway, while being able to look back at its past successes in many fields.

(Translated by Victor Kayfetz)

Nomination and Selection of the Nobel Laureates*

*compiled by Birgitta Lemmel***

The Royal Swedish Academy of Sciences *(with approximately 350 members) awards the Nobel Prizes in Physics and Chemistry and the Bank of Sweden Prize in Economic Sciences in Memory of Alfred Nobel (established in 1968).*

The Nobel Assembly at Karolinska Institutet *(with 50 members) awards the Nobel Prize in Physiology or Medicine.*

The Swedish Academy *(with 18 members) awards the Nobel Prize in Literature.*

These three institutions have special Nobel Committees of five members each — in the case of the Economics Prize, known as the Prize Committee — at their disposal for the preparatory work connected with the prize adjudication.

The Norwegian Nobel Committee*, whose five members are appointed by the Norwegian Parliament (Storting) awards the Nobel Peace Prize.*

One reason why the Prizes are awarded both in Sweden and Norway is that the two countries were united at the time Alfred Nobel wrote his will in 1895. The union between Norway and Sweden was dissolved in 1905, but this did not alter the relation among the Nobel institutions.

---

*The Special Regulations of the Statutes of the Nobel Foundation concerning nomination and selection are under revision and will not be finalized before early autumn 2001. For the latest version see www.nobelprize.org.

**Head of Information of the Nobel Foundation, 1986–1996.

Each year the respective Nobel Committees send individual invitations to thousands of scientists, members of academies and university professors in numerous countries, asking them to nominate candidates for the Nobel Prizes for the coming year. Those who are invited to submit nominations are chosen in such a way that as many countries and universities as possible are represented. These Prize nominations must reach the respective Committees before the first of February of the year for which the nomination is being made.

The nominations received by each Committee are then investigated with the help of specially appointed experts. When the Committees have made their selection among the nominated candidates and have presented their recommendations to the Prize-Awarding Institutions, a vote is taken for the final choice of Laureates. Prize decisions are announced immediately after the vote, which takes place in October. Eligibility to nominate candidates for the Nobel Prizes varies among the Prize-Awarding Institutions, as follows.

Physics and Chemistry

1. Swedish and foreign members of the Royal Swedish Academy of Sciences;
2. Members of the Nobel Committees for Physics and Chemistry;
3. Nobel Laureates in Physics and Chemistry;
4. Permanent and assistant professors in the sciences of Physics and Chemistry at the universities and institutes of technology of Sweden, Denmark, Finland, Iceland and Norway, and at Karolinska Institutet, Stockholm;
5. Holders of corresponding chairs in at least six universities or university colleges selected by the Academy of Sciences with a view to ensuring appropriate distribution over the different countries and their seats of learning; and
6. Other scientists from whom the Academy may see fit to invite proposals.

Decisions as to the selection of the teachers and scientists referred to in Paragraphs 5 and 6 above shall be taken each year before the end of September.

Physiology or Medicine

1. Members of the Nobel Assembly at Karolinska Institutet;
2. Swedish and foreign members of the medical class of the Royal Swedish Academy of Sciences;

3. Nobel Laureates in Physiology or Medicine;
4. Members of the Nobel Committee not qualified under Paragraph 1 above;
5. Holders of established posts as professors at the faculties of medicine in Sweden and holders of similar posts at the faculties of medicine or similar institutions in Denmark, Finland, Iceland and Norway;
6. Holders of similar posts at no fewer than six other faculties of medicine selected by the Assembly, with a view to ensuring the appropriate distribution of the task among various countries and their seats of learning; and
7. Practitioners of natural sciences whom the Assembly may otherwise see fit to approach.

Decisions concerning the selection of the persons appointed under Paragraphs 6 and 7 above are taken before the end of May each year on the recommendation of the Nobel Committee.

Literature

1. Members of the Swedish Academy and of other academies, institutions and societies which are similar to it in constitution and purpose;
2. Professors of literature and of linguistics at universities and university colleges;
3. Nobel Laureates in Literature; and
4. Presidents of those societies of authors that are representative of literary production in their respective countries.

Peace

1. Active and former members of the Norwegian Nobel Committee and the advisers appointed by the Norwegian Nobel Institute;
2. Members of the national assemblies and governments of the different states and members of the Inter-parliamentary Union;
3. Members of the International Court of Justice at the Hague and the International Court of Arbitration at the Hague;
4. Members of the Commission of the Permanent International Peace Bureau;
5. Members and associate members of the Institut de Droit International;
6. University professors of political science and jurisprudence, history and philosophy;
7. Nobel Peace Prize Laureates.

The Bank of Sweden Prize in Economic Sciences in Memory of Alfred Nobel

1. Swedish and foreign members of the Royal Swedish Academy of Sciences;
2. Members of the Prize Committee for the Bank of Sweden Prize in Economic Sciences in Memory of Alfred Nobel;
3. Prize Winners in Economic Sciences;
4. Permanent professors in relevant subjects at the universities and colleges in Sweden, Denmark, Finland, Iceland and Norway;
5. Holders of corresponding chairs in at least six universities or colleges selected for the relevant year by the Academy of Sciences with a view to ensuring an appropriate distribution among different countries and their seats of learning; and
6. Other scientists from whom the Academy may see fit to invite proposals.

Decisions as to the selection of the teachers and scientists referred to in Paragraphs 5 and 6 shall be taken each year before the end of the month of September.

PHYSICS

"...one part to the person who shall have made the most important discovery or invention within the field of physics..."

The Nobel Prize in Physics

*Erik B. Karlsson**

1. What is Physics?

Physics is considered to be the most basic of the natural sciences. It deals with the fundamental constituents of matter and their interactions as well as the nature of atoms and the build-up of molecules and condensed matter. It tries to give unified descriptions of the behavior of matter as well as of radiation, covering as many types of phenomena as possible. In some of its applications, it comes close to the classical areas of chemistry, and in others there is a clear connection to the phenomena traditionally studied by astronomers. Present trends are even pointing toward a closer approach of some areas of physics and microbiology.

Although chemistry and astronomy are clearly independent scientific disciplines, both use physics as a basis in the treatment of their respective problem areas, concepts and tools. To distinguish what is physics and chemistry in certain overlapping areas is often difficult. This has been illustrated several times in the history of the Nobel Prizes. Therefore, a few awards for chemistry will also be mentioned in the text that follows, particularly when they are closely connected to the works of the Physics Laureates themselves. As for astronomy, the situation is different since it has no Nobel Prizes of its own; it has therefore been natural from the start, to consider discoveries in astrophysics as possible candidates for Prizes in Physics.

---

*Professor Emeritus of Physics at Uppsala University, Sweden.

2. From Classical to Quantum Physics

In 1901, when the first Nobel Prizes were awarded, the classical areas of physics seemed to rest on a firm basis built by great 19th century physicists and chemists. Hamilton had formulated a very general description of the dynamics of rigid bodies as early as the 1830s. Carnot, Joule, Kelvin and Gibbs had developed thermodynamics to a high degree of perfection during the second half of the century.

Maxwell's famous equations had been accepted as a general description of electromagnetic phenomena and had been found to be also applicable to optical radiation and the radio waves recently discovered by Hertz.

Everything, including the wave phenomena, seemed to fit quite well into a picture built on mechanical motion of the constituents of matter manifesting itself in various macroscopic phenomena. Some observers in the late 19th century actually expressed the view that, what remained for physicists to do was only to fill in minor gaps in this seemingly well-established body of knowledge.

However, it would very soon turn out that this satisfaction with the state of physics was built on false premises. The turn of the century became a period of observations of phenomena that were completely unknown up to then, and radically new ideas on the theoretical basis of physics were formulated. It must be regarded as a historical coincidence, probably never foreseen by Alfred Nobel himself, that the Nobel Prize institution happened to be created just in time to enable the prizes to cover many of the outstanding contributions that opened new areas of physics in this period.

One of the unexpected phenomena during the last few years of the 19th century, was the discovery of X-rays by Wilhelm Conrad Röntgen in 1895, which was awarded the first Nobel Prize in Physics (1901). Another was the discovery of radioactivity by Antoine Henri Becquerel in 1896, and the continued study of the nature of this radiation by Marie and Pierre Curie. The origin of the X-rays was not immediately understood at the time, but it was realized that they indicated the existence of a hitherto concealed world of phenomena (although their practical usefulness for medical diagnosis was evident enough from the beginning). The work on radioactivity by Becquerel and the Curies was rewarded in 1903 (with one half to Becquerel and the other half shared by the Curies), and in combination with the additional work by Ernest Rutherford (who got the Chemistry Prize in 1908) it was understood that atoms, previously considered as more or less structureless objects, actually contained a very small but compact nucleus. Some atomic nuclei were found to be unstable and could emit the α, β or γ radiation observed. This was a revolutionary insight at the time, and it led in the end, through parallel work in other areas of physics, to the creation of the first useful picture of the structure of atoms.

In 1897, Joseph J. Thomson, who worked with rays emanating from the cathode in partly evacuated discharge tubes, identified the carriers of electric charge. He showed that these rays consisted of discrete particles, later called 'electrons'. He measured a value for the ratio between their mass and (negative) charge, and found that it was only a very small fraction of that expected for singly charged atoms. It was soon realized that these lightweight particles must be the building blocks that, together with the positively charged nuclei, make up all different kinds of atoms. Thomson received his Prize in 1906. By then, Philipp E. A. von Lenard had already been acknowledged the year before (1905) for elucidating other interesting properties of the cathodic rays, such as their ability to penetrate thin metal foils and produce fluorescence. Soon thereafter (in 1912) Robert A. Millikan made the first precision measurement of the electron charge with the oil-drop method, which led to a Physics Prize for him in 1923. Millikan was also rewarded for his works on the photoelectric effect.

In the beginning of the century, Maxwell's equations had already existed for several decades, but many questions remained unanswered: what kind of medium propagated electromagnetic radiation (including light) and what carriers of electric charges were responsible for light emission? Albert A. Michelson had developed an interferometric method, by which distances between objects could be measured as a number of wavelengths of light (or fractions thereof). This made comparison of lengths much more exact than what had been possible before. Many years later, the Bureau International de Poids et Mesures, Paris (BINP) defined the meter unit in terms of the number of wavelengths of a particular radiation instead of the meter prototype. Using such an interferometer, Michelson had also performed a famous experiment, together with E. W. Morley, from which it could be concluded that the velocity of light is independent of the relative motion of the light source and the observer. This fact refuted the earlier assumption of an ether as a medium for light propagation. Michelson received the Physics Prize in 1907.

The mechanisms for emission of light by carriers of electric charge was studied by Hendrik A. Lorentz, who was one of the first to apply Maxwell's equations to electric charges in matter. His theory could also be applied to the radiation caused by vibrations in atoms and it was in this context that it could be put to its first crucial test. As early as 1896 Pieter Zeeman, who was looking for possible effects of electric and magnetic fields on light, made an important discovery namely, that spectral lines from sodium in a flame were split up into several components when a strong magnetic field was applied. This phenomenon could be given a quite detailed interpretation by Lorentz's theory, as applied to vibrations of the recently identified electrons, and Lorentz and Zeeman shared the Physics Prize in 1902, i.e. even before Thomson's discovery was rewarded. Later, Johannes Stark demonstrated the direct effect of electric fields on the emission of light, by exposing beams of atoms ('anodic rays', consisting of atoms or molecules) to strong electric fields. He observed a complicated splitting of spectral lines as well as a Doppler shift depending on the velocities of the emitters. Stark received the 1919 Physics Prize.

With this background, it became possible to build detailed models for the atoms, objects that had existed as concepts ever since antiquity but were considered more or less structureless in classical physics. There existed already, since the middle of the previous century, a rich empirical material in the form of characteristic spectral lines emitted in the visible domain by different kinds of atoms, and to this was added the characteristic X-ray radiation discovered by Charles G. Barkla (Physics Prize in 1917, awarded in 1918), which after the clarification of the wave nature of this radiation and its diffraction by Max von Laue (Physics Prize in 1914), also became an important source of information on the internal structure of atoms.

Barkla's characteristic X-rays were secondary rays, specific for each element exposed to radiation from X-ray tubes (but independent of the chemical form of the samples). Karl Manne G. Siegbahn realized that measuring characteristic X-ray spectra of all the elements would show systematically, how successive electron shells are added when going from the light elements to the heavier ones. He designed highly accurate spectrometers for this purpose by which energy differences between different shells, as well as rules for radiative transitions between them, could be established. He received the Physics Prize in 1924 (awarded in 1925). However, it would turn out that a deeper understanding of the atomic structure required a much further departure from the habitual concepts of classical physics than anyone could have imagined.

Classical physics assumes continuity in motion as well as in the gain or loss of energy. Why then, do atoms send out radiations with sharp wavelengths? Here, a parallel line of development, also with its roots in late 19th-century physics, had given important clues for interpretation. Wilhelm

Wien studied the 'black-body' radiation from hot solid bodies (which in contrast to radiation from atoms in gases, has a continuous distribution of frequencies). Using classical electrodynamics, he derived an expression for the frequency distribution of this radiation and the shift of the maximum intensity wavelength, when the temperature of a black body is changed (the Wien displacement law, useful for instance in determining the temperature of the sun). He was awarded the Physics Prize in 1911.

However, Wien could not derive a distribution formula that agreed with experiments for both short and long wavelengths. The problem remained unexplained until Max K. E. L. Planck put forward his radically new idea that the radiated energy could only be emitted in quanta, i.e. portions that had a certain definite value, larger for the short wavelengths than for the long ones (equal to a constant h times the frequency for each quantum). This is considered to be the birth of quantum physics. Wien received the Physics Prize in 1911 and Planck some years later, in 1918 (awarded in 1919). Important verifications that light comes in the form of energy quanta came also through Albert Einstein's interpretation of the photoelectric effect (first observed in 1887 by Hertz) which also involved extensions of Planck's theories. Einstein received the Physics Prize for 1921 (awarded in 1922). The prize motivation cited also his other "services to theoretical physics," which will be referred to in another context.

Later experiments by James Franck and Gustav L. Hertz demonstrated the inverse of the photoelectric effect (i.e. that an electron that strikes an atom, must have a specific minimum energy to produce light quanta of a particular energy from it) and showed the general validity of Planck's expressions involving the constant h. Franck and Hertz shared the 1925 prize, awarded in 1926. At about the same time, Arthur H. Compton (who received one-half of the Physics Prize for 1927) studied the energy loss in X-ray photon scattering on material particles, and showed that X-ray quanta, whose energies are more than 10,000 times larger than those of light, also obey the same quantum rules. The other half was given to Charles T. R. Wilson (see later), whose device for observing high energy scattering events could be used for verification of Compton's predictions.

With the concept of energy quantization as a background, the stage was set for further ventures into the unknown world of microphysics. Like some other well-known physicists before him, Niels H. D. Bohr worked with a planetary picture of electrons circulating around the nucleus of an atom. He found that the sharp spectral lines emitted by the atoms could only be explained if the electrons were circulating in stationary orbits characterized by a quantized angular momentum (integer units of Planck's constant h divided by 2π) and that the emitted frequencies ν corresponded to emission of radiation with energy $h\nu$ equal to the difference between quantized energy states of the electrons. His suggestion indicated a still more radical departure from classical physics than Planck's hypothesis. Although it could only explain some of the simplest features of optical spectra in its original form, it was soon accepted that Bohr's approach must be a correct starting point, and he received the Physics Prize in 1922.

It turned out that a deeper discussion of the properties of radiation and matter (until then considered as forming two completely different categories), was necessary for further progress in the theoretical description of the microworld. In 1923 Prince Louis-Victor P. R. de Broglie proposed that material particles may also show wave properties, now that electromagnetic radiation had been shown to display particle aspects in the form of photons. He developed mathematical expressions for this dualistic behaviour, including what has later been called the "de Broglie wavelength" of a moving particle. Early experiments by Clinton J. Davisson had indicated that electrons could actually show reflection effects similar to that of waves hitting a crystal and these experiments were now repeated, verifying the associated wavelength predicted by de Broglie. Somewhat later, George P. Thomson (son of J. J. Thomson) made much improved experiments on higher energy electrons penetrating thin metal foils which showed very clear diffraction effects. de Broglie was rewarded for his theories in 1929 and Davisson and Thomson shared later the 1937 Physics Prize.

What remained was the formulation of a new, consistent theory that would replace classical mechanics, valid for atomic phenomena and their associated radiations. The years 1924–1926 was a period of intense development in this area. Erwin Schrödinger built further on the ideas of de Broglie and wrote a fundamental paper on "Quantization as an eigenvalue problem" early in 1926. He created what has been called 'wave mechanics'. But the year before that, Werner K. Heisenberg had already started on a mathematically different approach, called 'matrix mechanics', by which he arrived at equivalent results (as was later shown by Schrödinger). Schrödinger's and Heisenberg's new quantum mechanics meant a fundamental departure from the intuitive picture of classical orbits for atomic objects, and implied also that there are natural limitations on the accuracy by which certain quantities can be measured simultaneously (Heisenberg's uncertainty relations).

Heisenberg was rewarded by the Physics Prize for 1932 (awarded 1933) for the development of quantum mechanics, while Schrödinger shared the

Prize one year later (1933) with Paul A. M. Dirac. Schrödinger's and Heisenberg's quantum mechanics was valid for the relatively low velocities and energies associated with the 'orbital' motion of valence electrons in atoms, but their equations did not satisfy the requirements set by Einstein's rules for fast moving particles (to be mentioned later). Dirac constructed a modified formalism which took into account effects of Einstein's special relativity, and showed that such a theory not only contained terms corresponding to the intrinsic spinning of electrons (and therefore explaining their own intrinsic magnetic moment and the fine structure observed in atomic spectra), but also predicted the existence of a completely new kind of particles, the so-called antiparticles with identical masses but opposite charge. The first antiparticle to be discovered, that of the electron, was observed in 1932 by Carl D. Anderson and was given the name 'positron' (one-half of the Physics Prize for 1936).

Other important contributions to the development of quantum theory have been distinguished by Nobel Prizes in later years. Max Born, Heisenberg's supervisor in the early twenties, made important contributions to its mathematical formulation and physical interpretation. He received one-half of the Physics Prize for 1954 for his work on the statistical interpretation of the wave function. Wolfgang Pauli formulated his exclusion principle (which states that there can be only one electron in each quantum state) already on the basis of Bohr's old quantum theory. This principle was later found to be associated with the symmetry of wave functions for particles of half-integer spins in general, distinguishing what is now called fermions from the bosonic particles whose spins are integer multiples of $h/2\pi$. The exclusion principle has deep consequences in many areas of physics and Pauli received the Nobel Prize in Physics in 1945.

The study of electron spins would continue to open up new horizons in physics. Precision methods for determining the magnetic moments of spinning particles were developed during the thirties and forties for atoms as well as nuclei (by Stern, Rabi, Bloch and Purcell, see later sections) and in 1947 they had reached such a precision, that Polykarp Kusch could state that the magnetic moment of an electron did not have exactly the value predicted by Dirac, but differed from it by a small amount. At about the same time, Willis E. Lamb worked on a similar problem of electron spins interacting with electromagnetic fields, by studying the fine structure of optical radiation from hydrogen with very high resolution radio frequency resonance methods. He found that the fine structure splitting also did not have exactly the Dirac value, but differed from it by a significant amount. These results stimulated a reconsideration of the basic concepts behind the application of quantum theory to electromagnetism, a field that had been started by Dirac, Heisenberg and Pauli but still suffered from several insufficiencies. Kusch and Lamb were each awarded half the Physics Prize in 1955.

In quantum electrodynamics (QED for short), charged particles interact through the interchange of virtual photons, as described by quantum perturbation theory. The older versions involved only single photon exchange, but Sin–Itiro Tomonaga, Julian Schwinger and Richard P. Feynman realized that the situation is actually much more complicated, since electron-electron scattering may involve several photon exchanges. A 'naked' point charge does not exist in their picture; it always produces a cloud of virtual particle-antiparticle pairs around itself, such that its effective magnetic moment is changed and the Coulomb potential is modified at short distances. Calculations starting from this picture have reproduced the experimental data by Kusch and Lamb to an astonishing degree of accuracy and modern QED is now considered to be the most exact theory in existence. Tomonaga, Schwinger and Feynman shared the Physics Prize in 1965.

This progress in QED turned out to be of the greatest importance also for the description of phenomena at higher energies. The notion of pair production from a 'vacuum' state of a quantized field (both as a virtual process and as a real materialization of particles), is also a central building block in the modern field theory of strong interactions, quantum chromodynamics (QCD).

Another basic aspect of quantum mechanics and quantum field theory is the symmetries of wave functions and fields. The symmetry properties under exchange of identical particles lie behind Pauli's exclusion principle mentioned above, but symmetries with respect to spatial transformations have turned out to play an equally important role. In 1956, Tsung-Dao Lee and Chen Ning Yang pointed out, that physical interactions may not always be symmetric with respect to reflection in a mirror (that is, they may be different as seen in a left-handed and a right-handed coordinate system). This means that the wave function property called 'parity', denoted by the symbol 'P', is not conserved when the system is exposed to such an interaction and the mirror reflection property may be changed. Lee's and Yang's work was the starting point for an intense search for such effects and it was shown soon afterwards that the β decay and the $\pi \rightarrow \mu$ decay, which are both caused by the so-called 'weak interaction' are not parity-conserving (see more below). Lee and Yang were jointly awarded the Physics Prize in 1957.

Other symmetries in quantum mechanics are connected with the replacement of a particle with its antiparticle, called charge conjugation (symbolized by 'C'). In the situations discussed by Lee and Yang it was found that although parity was not conserved in the radioactive transformations there was still a symmetry in the sense that particles and antiparticles broke parity in exactly opposite ways and that therefore the combined operation 'C' × 'P' still gave results which preserved symmetry. But it did not last long before James W. Cronin and Val L. Fitch found a decay mode among the 'K mesons' that violated even this principle, although only to a small extent. Cronin and Fitch made their discovery in 1964 and were jointly awarded the Physics Prize in 1980. The consequences of their result (which include questions about the symmetry of natural processes under reversal of time, called 'T') are still discussed today and touch some of the deepest foundations of theoretical physics, because the 'P' × 'C' × 'T' symmetry is expected always to hold.

The electromagnetic field is known to have another property, called 'gauge symmetry', which means that the field equations keep their form even if the electromagnetic potentials are multiplied with certain quantum mechanical phase factors, or 'gauges'. It was not self-evident that the 'weak' interaction should have this property, but it was a guiding principle in the work by Sheldon L. Glashow, Abdus Salam, and Steven Weinberg in the late 1960s, when they formulated a theory that described the weak and the electromagnetic interaction on the same basis. They were jointly awarded the Physics Prize in 1979 for this unified description and, in particular, for their prediction of a particular kind of weak interaction mediated by 'neutral currents', which had been found recently in experiments.

The last Physics Prize (1999) in the 20th century was jointly awarded to Gerardus 't Hooft and Martinus J. G. Veltman. They showed the way to renormalize the 'electro-weak' theory, which was necessary to remove terms that tended to infinity in quantum mechanical calculations (just as QED had earlier solved a similar problem for the Coulomb interaction). Their work allowed detailed calculations of weak interaction contributions to particle interactions in general, proving the utility of theories based on gauge invariance for all kinds of basic physical interactions.

Quantum mechanics and its extensions to quantum field theories is one of the great achievements of the 20th century. This sketch of the route from classical physics to modern quantum physics, has led us a long way toward a fundamental and unified description of the different particles and forces in nature, but much remains to be done and the goal is still far ahead. It still remains, for instance, to 'unify' the electro-weak force with the 'strong' nuclear force and with gravity. But here, it should also be pointed out that the quantum description of the microworld has another main application: the calculation of chemical properties of molecular systems (sometimes extended to biomolecules) and of the structure of condensed matter, branches that have been distinguished with several prizes, in physics as well as in chemistry.

3. Microcosmos and Macrocosmos

'From Classical to Quantum Physics', took us on a trip from the phenomena of the macroscopic world as we meet it in our daily experience, to the quantum world of atoms, electrons and nuclei. With the atoms as starting point, the further penetration into the subatomic microworld and its smallest known constituents will now be illustrated by the works of other Nobel Laureates.

It was realized, already in the first half of the 20th century, that such a further journey into the microcosmos of new particles and interactions would also be necessary for understanding the composition and evolution histories of the very large structures of our universe, the 'macrocosmos'. At the present stage elementary particle physics, astrophysics, and cosmology are strongly tied together, as several examples presented here will show.

Another link connecting the smallest and the largest objects in our universe is Albert Einstein's theories of relativity. Einstein first developed his special theory of relativity in 1905, which expresses the mass-energy relationship $E = mc^2$. Then, in the next decade, he continued with his theory of general relativity, which connects gravitational forces to the structure of space and time. Calculations of effective masses for high energy particles, energy transformations in radioactive decay as well as Dirac's predictions that antiparticles may exist, are all based on his special theory of relativity. The general theory is the basis for calculations of large scale motions in the universe, including discussions of the properties of black holes. Einstein received the Nobel Prize in Physics in 1921 (awarded in 1922), motivated by work on the photo-electric effect which demonstrated the particle aspects of light.

The works by Becquerel, the Curies, and Rutherford gave rise to new questions: What was the source of energy in the radioactive nuclei that could sustain the emission of α, β and γ radiation over very long time intervals, as observed for some of them, and what were the heavy α particles and the nuclei themselves actually composed of? The first of these problems (which seemed to violate the law of conservation of energy, one of the most important principles of physics) found its solution in the transmutation theory, formulated by Rutherford and Frederick Soddy (Chemistry Prize for 1921, awarded in 1922). They followed in detail several different series of radioactive decay and compared the energy emitted with the mass differences between 'parent' and 'daughter' nuclei. It was also found that nuclei belonging to the same chemical element could have different masses; such different species were called 'isotopes'. A Chemistry Prize was given in 1922 to Francis W. Aston for his mass-spectroscopic separation of a large number of isotopes of non-radioactive elements. Marie Curie had by then already received a second Nobel Prize (this time in Chemistry in 1911), for her discoveries of the chemical elements radium and polonium.

All isotopic masses were found to be nearly equal to multiples of the mass of the proton, a particle also first seen by Rutherford when he irradiated nitrogen nuclei with α particles. But the different isotopes could not possibly be made up entirely of protons since each particular chemical element must have one single value for the total nuclear charge. Protons were actually found to make up less than half of the nuclear mass, which meant that some neutral constituents had to be present in the nuclei. James Chadwick first found conclusive evidence for such particles, the neutrons, when he studied nuclear reactions in 1932. He received the Physics Prize in 1935.

Soon after Chadwick's discovery, neutrons were put to work by Enrico Fermi and others as a means to induce nuclear reactions that could produce new 'artificial' radioactivity. Fermi found that the probability for neutron-induced reactions (which do not involve element transformations), increased when the neutrons were slowed down and that this worked equally well for heavy elements as for light ones, in contrast to charge-particle induced reactions. He received the Physics Prize in 1938.

With neutrons and protons as the basic building blocks of atomic nuclei, the branch of 'nuclear physics' could be established and several of its major achievements were distinguished by Nobel prizes. Ernest O. Lawrence, who received the Physics Prize in 1939, built the first cyclotron in which acceleration took place by successively adding small amounts of energy to particles circulating in a magnetic field. With these machines, he was able to accelerate charged nuclear particles to such high energies that they could induce nuclear reactions and he obtained important new results. Sir John D. Cockcroft and Ernest T. S. Walton instead, accelerated particles by direct application of very high electrostatic voltages and were rewarded for their studies of transmutation of elements in 1951.

Otto Stern received the Physics Prize in 1943 (awarded in 1944), for his experimental methods of studying magnetic properties of nuclei, in particular for measuring the magnetic moment of the proton itself. Isidor I. Rabi increased the accuracy of magnetic moment determinations for nuclei by more than two orders of magnitude, with his radio frequency resonance technique, for which he was awarded the Physics Prize for 1944. Magnetic properties of nuclei provide important information for understanding details in the build-up of the nuclei from protons and neutrons. Later, in the second half of the century, several theoreticians were rewarded for their work on the theoretical modelling of this complex many-body system: Eugene P. Wigner (one-half of the prize), Maria Goeppert-Mayer (one-fourth) and J. Hans D. Jensen (one-fourth) in 1963 and Aage N. Bohr, Ben R. Mottelson and L. James Rainwater in 1975. We will come back to these works under the heading 'From Simple to Complex Systems'.

As early as 1912, it was found by Victor F. Hess (awarded half the Prize in 1936 with the other half to Carl D. Anderson) that highly penetrating radiation is also reaching us continuously from outer space. This 'cosmic radiation' was first detected by ionization chambers and later by Wilson's cloud chamber referred to earlier. Properties of particles in the cosmic radiation could be inferred from the curved particle tracks produced when a strong magnetic field was applied. It was in this way that C. D. Anderson discovered the positron. Anderson and Patrick M. S. Blackett showed that electron positron pairs could be produced by γ rays (which needed a photon energy equal to at least $2m_ec^2$) and that electrons and positrons could annihilate, producing γ rays as they disappeared. Blackett received the Physics Prize in 1948 for his further development of the cloud chamber and the discoveries made with it.

Although accelerators were further developed, cosmic radiation continued for a couple of decades to be the main source of very energetic particles (and still surpasses the most powerful accelerators on earth in this aspect, although with extremely low intensities), and it provided the first glimpses of a completely unknown subnuclear world. A new kind of particles, called mesons, was spotted in 1937, having masses approximately 200 times that of electrons (but 10 times lighter than protons). In 1946, Cecil F. Powell clarified the situation by showing that there were actually more than one kind of such particles present. One of them, the 'π meson', decays into the other one, the 'μ meson'. Powell was awarded the Physics Prize in 1950.

By that time, theoreticians had already been speculating about the forces that keep protons and neutrons together in nuclei. Hideki Yukawa suggested in 1935, that this 'strong' force should be carried by an exchange particle, just as the electromagnetic force was assumed to be carried by an exchange of virtual photons in the new quantum field theory. Yukawa maintained that such a particle must have a mass of about 200 electron masses in order to explain the short range of the strong forces found in experiments. Powell's π meson was found to have the right properties to act as a 'Yukawa particle'. The μ particle, on the other hand, turned out to have a completely different character (and its name was later changed from 'μ meson' to 'muon'). Yukawa received the Physics Prize in 1949. Although later progress has shown that the strong force mechanism is more complex than what Yukawa pictured it to be, he must still be considered the first one who led the thoughts on force carriers in this fruitful direction.

More new particles were discovered in the 1950s, in cosmic radiation as well as in collisions with accelerated particles. By the end of the 50s, accelerators could reach energies of several GeV (10^9 electron volts) which meant that pairs of particles, with masses equal to the proton mass, could be created by energy-to-mass conversion. This was the method used by the team of Owen Chamberlain and Emilio Segrè when they first identified and studied the antiproton in 1955 (they shared the Physics Prize for 1959). High energy accelerators also allowed more detailed studies of the structures of protons and neutrons than before, and Robert Hofstadter was able to distinguish details of the electromagnetic structure of the nucleons by observing how they scattered electrons of very high energy. He was rewarded with half the Physics Prize for 1961.

One after another, new mesons with their respective antiparticles appeared, as tracks on photographic plates or in electronic particle detectors. The existence of the 'neutrino' predicted on theoretical grounds by Pauli already as early as the 1930s, was established. The first direct experimental evidence for the neutrino was provided by C. L. Cowan and Frederick Reines in 1957, but it was not until 1995 that this discovery was awarded with one-half the Nobel Prize (Cowan had died in 1984). The neutrino is a participant in processes involving the 'weak' interaction (such as β decay and π meson decay to muons) and, as the intensity of particle beams increased, it became possible to produce secondary beams of neutrinos from accelerators. Leon M. Lederman, Melvin Schwartz and Jack Steinberger developed this method in the 1960s and demonstrated that the neutrinos accompanying μ emission in π decay were not identical to those associated with electrons in β decay; they were two different particles, ν_μ and ν_e.

Physicists could now start to distinguish some order among the particles: the electron (e), the muon (μ), the electron neutrino (ν_e), the muon neutrino (ν_μ) and their antiparticles were found to belong to one class, called 'leptons'. They did not interact by the 'strong' nuclear force, which on the other hand, characterized the protons, neutrons, mesons and hyperons (a set of particles heavier than the protons). The lepton class was extended later in the 1970s when Martin L. Perl and his team discovered the τ lepton, a heavier relative to the electron and the muon. Perl shared the Physics Prize in 1995 with Reines.

All the leptons are still considered to be truly fundamental, i.e. point-like and without internal structure, but for the protons, etc, this is no longer true. Murray Gell-Mann and others managed to classify the strongly interacting particles (called 'hadrons') into groups with common relationships and ways of interaction. Gell-Mann received the Physics Prize in 1969. His systematics was based on the assumption that they were all built up from more elementary constituents, called 'quarks'. The real proof that nucleons were built up from quark-like objects came through the works of Jerome I. Friedman, Henry W. Kendall and Richard E. Taylor. They 'saw' hard grains inside these objects when they studied how electrons (of still higher energy than Hofstadter could use earlier) scattered inelastically on them. They shared the Physics Prize in 1990.

It was understood that all strongly interacting particles are built up by quarks. In the middle of the 1970s a very short-lived particle, discovered independently by the teams of Burton Richter and Samuel C. C. Ting, was found to contain a so far, unknown type of quark which was given the name 'charm'. This quark was a missing link in the systematics of the elementary particles and Burton and Ting shared the Physics Prize in 1976. The present standard model of particle physics sorts the particles into three families, with two quarks (and their antiparticles) and two leptons in each: the 'up' and 'down' quarks, the electron and the electron-neutrino in the first; the 'strange' and the 'charm' quark, the muon and the muon neutrino in the second; the 'top' and the 'bottom' quark, the tauon and the tau neutrino in the third. The force carriers for the combined electro-weak interaction are the photon, the Z-particle and the W-bosons, and for the strong interaction between quarks the so-called gluons.

In 1983, the existence of the W- and Z-particles was proven by Carlo Rubbia's team which used a new proton-antiproton collider with sufficient energy for production of these very heavy particles. Rubbia shared the 1984 physics prize with Simon van der Meer who had made decisive contributions to the construction of this collider by his invention of 'stochastic cooling' of particles. There are speculations that additional particles may be produced at energies higher than those attainable with the present accelerators, but no experimental evidence has been produced so far.

Cosmology is the science that deals with the structure and evolution of our universe and the large-scale objects in it. Its models are based on the properties of the known fundamental particles and their interactions as well as the properties of space–time and gravitation. The 'big-bang' model describes a possible scenario for the early evolution of the universe. One of its predictions was experimentally verified when Arno A. Penzias and Robert W. Wilson discovered the cosmic microwave radiation background in 1960. They shared one-half of the Physics Prize for 1978. This radiation is an afterglow of the violent processes assumed to have occurred in the early stages of the big bang. Its equilibrium temperature is 3 kelvin at the present age of the universe. It is almost uniform when observed in different directions; the small deviations from isotropy are now being investigated and will tell us more about the earliest history of our universe.

Outer space has been likened to a large arena for particle interactions where extreme conditions, not attainable in a laboratory, are spontaneously created. Particles may be accelerated to higher energies than in any accelerator on earth, nuclear fusion reactions proliferate in the interior of stars, and gravitation can compress particle systems to extremely high densities. Hans A. Bethe first described the hydrogen and carbon cycles, in which energy is liberated in stars by the fusion of protons into helium nuclei. For this achievement he received the Physics Prize in 1967.

Subramanyan Chandrasekhar described theoretically the evolution of stars, in particular those ending up as 'white dwarfs'. Under certain conditions the end product may also be a 'neutron star', an extremely compact object, where all protons have been converted into neutrons. In supernova explosions, the heavy elements created during stellar evolution are spread out into space. The details of some of the most important nuclear reactions in stars and heavy element formation were elucidated by William A. Fowler both in theory and in experiments using accelerators. Fowler and Chandrasekhar received one-half each of the 1983 Physics Prize.

Visible light and cosmic background radiation are not the only forms of electromagnetic waves that reach us from outer space. At longer wavelengths, radio astronomy provides information on astronomical objects not obtainable by optical spectroscopy. Sir Martin Ryle developed the method where signals from several separated telescopes are combined in order to increase the resolution in the radio source maps of the sky. Antony Hewish and his group made an unexpected discovery in 1964 using Ryle's telescopes: radio frequency pulses were emitted with very well-defined repetition rates by some unknown objects called pulsars. These were soon identified as neutron stars, acting like fast rotating lighthouses emitting radiowaves because they are also strong magnets. Ryle and Hewish shared the Physics Prize in 1974.

By 1974, pulsar search was already routine among radio astronomers, but a new surprise came in the summer of the same year when Russell A. Hulse and Joseph H. Taylor, Jr. noticed periodic modulations in the pulse frequencies of a newly discovered pulsar, called PSR 1913 + 16. It was the first double pulsar detected, so named because the emitting neutron star happened to be one of the components of a close double star system, with the other component of about equal size. This system has provided, by observation over more than 20 years, the first concrete evidence for gravitational radiation. The decrease of its rotational frequency is in close agreement with the predictions based on Einstein's theory, for losses caused by this kind of radiation. Hulse and Taylor shared the Physics Prize in 1993. However, the direct detection of gravitational radiation on earth still has to be made.

4. From Simple to Complex Systems

If all the properties of the elementary particles as well as the forces that may act between them were known in every detail, would it then be possible to predict the behavior of all systems composed of such particles? The search for the ultimate building blocks of nature and of the proper theoretical description of their interactions (on the macro as well as the micro scale), has partly been motivated by such a reductionistic program. All scientists would not agree that such a synthesis is possible even in principle. But even if it were true, the calculations of complex system behaviour would very soon be impossible to handle when the number of particles and interactions in the system is increased. Complex multi-particle systems are therefore described in terms of simplified models, where only the most essential features of their particle compositions and interactions are used as starting points. Quite often, it is observed that complex systems develop features called 'emergent properties', not straightforwardly predictable from the basic interactions between their constituents.

4.1. *Atomic Nuclei*

The first complex systems from the reductionist's point of view are the nucleons, i.e. neutrons and protons composed of quarks and gluons. The second is the atomic nuclei, which to a first approximation are composed of separate nucleons. The first advanced model of nuclear structure was the nuclear shell model, put up by the end of the 1940s by Maria Goeppert-Mayer and Johannes D. Jensen who realized that at least for nuclei with nearly spherical shape, the outer nucleons fill up energy levels like electrons in atoms. However, the order is different; it is determined by another common potential and by the specific strong spin-orbit coupling of the nuclear forces. Their model explains why nuclei with so-called 'magic numbers' of protons or neutrons are particularly stable. They shared the Physics Prize in 1963 together with Eugene Wigner, who had formulated fundamental symmetry principles important in both nuclear and particle physics.

Nuclei with nucleon numbers far from the magic ones are not spherical. Niels Bohr had already worked with a liquid drop model for such deformed nuclei which may take ellipsoidal shapes, and in 1939 it was found that excitation of certain strongly deformed nuclei could lead to nuclear fission, i.e. the breakup of such nuclei into two heavy fragments. Otto Hahn received the Chemistry Prize in 1944 (awarded in 1945) for the discovery of this new process. The non-spherical shape of deformed nuclei allows new collective, rotational degrees of freedom, as do also the collective vibrations of nucleons. Models describing such excitations of the nuclei were developed by James Rainwater, Aage Bohr (son of Niels Bohr) and Ben Mottelson, who jointly received the Physics Prize in 1975.

The nuclear models mentioned above, were based not only on general, guiding principles, but also on the steadily increasing information from nuclear spectroscopy. Harold C. Urey discovered deuterium, a heavy isotope of hydrogen, for which he was awarded the Chemistry Prize in 1934. Fermi, Lawrence, Cockcroft, and Walton mentioned in the previous section developed methods for the production of unstable nuclear isotopes. For their extension of the nuclear isotope chart to the heaviest elements, Edwin M. McMillan and Glenn T. Seaborg were awarded, again with a Chemistry Prize (in 1951). In 1954, Walther Bothe received one-half of the Physics Prize and the other half was awarded to Max Born, mentioned earlier. Bothe developed the coincidence method, which allowed spectroscopists to select generically related sequences of nuclear radiation from the decay of nuclei. This turned out to be important, particularly for the study of excited states of nuclei and their electromagnetic properties.

4.2. *Atoms*

The electronic shells of the atoms, when considered as many-body systems, are easier to handle than the nuclei (which actually contain not only protons and neutrons but also more of other, short-lived 'virtual' particles than the atoms). This is due to the weakness and simplicity of the electromagnetic forces as compared to the 'strong' forces holding the nuclei together. With the quantum mechanics developed by Schrödinger, Heisenberg, and Pauli, and the relativistic extensions by Dirac, the main properties of the atomic electrons could be reasonably well described. However, a long standing problem has remained, namely to solve the mathematical problems connected with the mutual interactions between the electrons after the dominating attraction by the positive nuclei has been taken into account. One aspect of this was addressed in the work by one of the most recent Chemistry Laureates (1998), Walter Kohn. He developed the 'density functional' method which is applicable to free atoms as well as to electrons in molecules and solids.

At the beginning of the 20th century, the periodic table of elements was not yet complete. The early history of the Nobel Prizes includes the discoveries of some of the then missing elements. Lord Raleigh (John William Strutt) noticed anomalies in the relative atomic masses when oxygen and nitrogen samples were taken directly from the air that surrounds us, instead of separating them from chemical compounds. He concluded that the atmosphere must contain a so far unknown constituent, which was the element argon with atomic mass 20. He was awarded the Physics Prize in 1904, the same year that Sir William Ramsay obtained the Chemistry Prize for isolating the element helium.

In the second half of the 20th century, there has been a spectacular development of atomic spectroscopy and the precision by which one can measure the transitions between atomic or molecular states that fall in the microwave and optical range. Alfred Kastler (who received the Physics Prize in 1966) and his co-workers, showed in the 1950s that electrons in atoms can be put into selected excited substates by the use of polarized light. After radiative decay, this can also lead to an orientation of the spins of ground-state atoms. The subsequent induction of radio frequency transitions opened possibilities to measure properties of the quantized states of electrons in atoms in much greater detail than before. A parallel line of development led to the invention of masers and lasers, which are based on the 'amplification of stimulated emission of radiation' in strong microwave and optical (light) fields, respectively (effects which in principle would have been predictable from Einstein's equations formulated in 1917 but were not discussed in practical terms until early in the 1950s).

Charles H. Townes developed the first maser in 1958. Theoretical work on the maser principle was made by Nikolay G. Basov and Aleksandr M. Prokhorov. The first maser used a stimulated transition in the ammonia molecule. It emitted an intense microwave radiation, which unlike that of natural emitters, was coherent (i.e. with all photons in phase). Its frequency sharpness soon made it an important tool in technology, for time-keeping and other purposes. Townes received half the Physics Prize for 1964 and Basov and Prokhorov shared the other half.

For radiation in the optical range, lasers were later developed in several laboratories. Nicolaas Bloembergen and Arthur L. Schawlow were distinguished in 1981 for their work on precision laser spectroscopies of atoms and molecules. The other half of that year's prize was awarded to Kai M. Siegbahn (son of Manne Siegbahn), who developed another high-precision method for atomic and molecular spectroscopy based on electrons emitted from inner electron shells when hit by X-rays with very well-defined energy. His photo- and Auger-electron spectroscopy is used as an analytical tool in several other areas of physics and chemistry.

The controlled interplay between atomic electrons and electromagnetic fields has continued to provide ever more detailed information about the structure of electronic states in atoms. Norman F. Ramsey developed precision methods based on the response to external radio frequency signals by free atoms in atomic beams and Wolfgang Paul invented atomic 'traps', built by combinations of electric and magnetic fields acting over the sample volumes. Hans G. Dehmelt's group was the first to isolate single particles (positrons) as well as single atoms in such traps. For the first time, experimenters could 'communicate' with individual atoms by microwave and laser signals. This enabled the study of new aspects of quantum mechanical behavior as well as further increased precision in atomic properties and the setting of time standards. Paul and Dehmelt received the 1989 Physics Prize and the other half was awarded to Ramsey.

The latest step in this development has involved the slowing down of the motion of atoms in traps to such an extent that it would correspond to micro-kelvin temperatures, had they been in thermal equilibrium in a gas. This is done by exposing them to 'laser cooling' through a set of ingenious schemes designed and carried out in practice by Steven Chu, Claude Cohen-Tannoudji and William D. Phillips, whose research groups manipulated atoms by collisions with laser photons. Their work, which was recognized by the 1997 Physics Prize, promises important applications in general measurement technology in addition to a still more increased precision in the determination of atomic quantities.

4.3. *Molecules and Plasmas*

Molecules are composed of atoms. They form the next level of complexity when considered as many-body systems. But molecular phenomena have traditionally been viewed as a branch of chemistry (as exemplified by the Chemistry Prize in 1936 to Petrus J. W. Debye), and have only rarely been in the focus for Nobel Prizes in Physics. One exception is the recognition of the work by Johannes Diderik van der Waals, who formulated an equation of state for molecules in a gas taking into account the mutual interaction between the molecules as well as the reduction of the free volume due to their finite size. van der Waals' equation has been an important starting point for the description of the condensation of gases into liquids. He received the 1910 Physics Prize. Jean B. Perrin studied the motion of small particles suspended in water and received the 1926 Physics Prize. His studies allowed a confirmation of Einstein's statistical theory of Brownian motion as well as of the laws governing the equilibrium of suspended particles under the influence of gravity.

In 1930, Sir C. Venkata Raman received the Physics Prize for his observations that light scattered from molecules contained components which were shifted in frequency with respect to the infalling monochromatic light. These shifts are caused by the molecules' gain or loss of characteristic amounts of energy when they change their rotational or vibrational motion. Raman spectroscopy soon became an important source of information on molecular structure and dynamics.

A plasma is a gaseous state of matter in which the atoms or molecules are strongly ionized. Mutual electromagnetic forces, both between the positive ions themselves and between the ions and the free electrons, are then playing dominant roles, which adds to the complexity as compared to the situation in neutral atomic or molecular gases. Hannes Alfvén demonstrated in the 1940s that a new type of collective motion, called 'magneto-hydrodynamical waves' can arise in such systems. These waves play a crucial role for the behavior of plasmas, in the laboratory as well as in the earth's atmosphere and in cosmos. Alfvén received half of the 1970 Physics Prize.

4.4. *Condensed Matter*

Crystals are characterized by a regular arrangement of atoms. Relatively soon after the discovery of the X-rays, it was realized by Max von Laue that such rays were diffracted when passing through crystalline solids, like light passing an optical grating. This effect is related to the fact that the wavelength of common X-ray sources happens to coincide with typical distances between atoms in these materials. It was first used systematically by Sir William Henry Bragg and William Lawrence Bragg (father and son) to measure interatomic distances and to analyse the geometrical arrangement of atoms in simple crystals. For their pioneering work on X-ray crystallography (which has later been developed to a high degree of sophistication), they received the Nobel Prize in Physics; Laue in 1914 and the Braggs in 1915.

The crystalline structure is the most stable of the different ways in which atoms can be organized to form a certain solid at the prevalent temperature and pressure conditions. In the 1930s Percy W. Bridgman invented devices by which very high pressures could be applied to different solid materials and studied changes in their crystalline, electric, magnetic and thermal properties. Many crystals undergo phase transitions under such extreme circumstances, with abrupt changes in the geometrical arrangements of their atoms at certain well-defined pressures. Bridgman received the Physics Prize in 1946 for his discoveries in the field of high pressure physics.

Low-energy neutrons became available in large numbers to the experimenters through the development of fission reactors in the 1940s. It was found that these neutrons, like X-rays, were useful for crystal structure determinations because their associated de Broglie wavelengths also fall in the range of typical interatomic distances in solids. Clifford G. Shull contributed strongly to the development of the neutron diffraction technique for crystal structure determination, and showed also that the regular arrangement of magnetic moments on atoms in ordered magnetic materials can give rise to neutron diffraction patterns, providing a new powerful tool for magnetic structure determination.

Shull was rewarded with the Physics Prize in 1994, together with Bertram N. Brockhouse, who specialized in another aspect of neutron scattering on condensed material: the small energy losses resulting when neutrons excite vibrational modes (phonons) in a crystalline lattice. For this purpose, Brockhouse developed the 3-axis neutron spectrometer, by which complete dispersion curves (phonon energies as function of wave vectors) could be obtained. Similar curves could be recorded for vibrations in magnetic lattices (the magnon modes).

John H. Van Vleck made significant contributions to the theory of magnetism in condensed matter in the years following the creation of quantum mechanics. He calculated the effects of chemical binding on the paramagnetic atoms and explained the effects of temperature and applied magnetic fields on their magnetism. In particular, he developed the theory of crystal field effects on the magnetism of transition metal compounds, which has been of great importance for understanding the function of active centers in compounds for laser physics as well as in biomolecules. He shared the Physics Prize in 1977 with Philip W. Anderson and Sir Nevill F. Mott (see below).

Magnetic atoms can have their moments all ordered in the same direction in each domain (ferromagnetism), with alternating 'up' and 'down' moments of the same size (simple antiferromagnets) or with more complicated patterns including different magnetic sublattices (ferrimagnets, etc). Louis E. F. Néel introduced basic models to describe antiferromagnetic and ferrimagnetic materials, which are important components in many solid state devices. They have been extensively studied by the aforementioned neutron diffraction techniques. Néel obtained one-half of the Physics Prize in 1970.

The geometric ordering of atoms in crystalline solids as well as the different kinds of magnetic order, are examples of general ordering phenomena in nature when systems find an energetically favourable arrangement by choosing a certain state of symmetry. The critical phenomena, which occur when transitions between states of different symmetry are approached (for instance when temperature is changed), have a high degree of universality for different types of transitions, including the magnetic ones. Kenneth G. Wilson, who received the Physics Prize in 1982, developed the so-called renormalization theory for critical phenomena in connection with phase transitions, a theory which has also found application in certain field theories of particle physics.

Liquid crystals form a specific class of materials that show many interesting features, from the point of view of fundamental interactions in condensed matter as well as for technical applications. Pierre-Gilles de Gennes developed the theory for the behavior of liquid crystals and their transitions between different ordered phases (nematic, smectic, etc). He also used statistical mechanics to describe the arrangements and dynamics of polymer chains, thereby showing that methods developed for ordering phenomena in simple systems can be generalized to the complex ones occurring in 'soft condensed matter'. For this, he received the Physics Prize in 1991.

Another specific form of liquid that has received strong attention is the liquid helium. At normal pressures, this substance remains liquid down to the lowest temperatures attainable. It also shows large isotope effects, since ^{4}He condenses to liquid at 4.2 K, while the more rare isotope ^{3}He remains in gaseous form down to 3.2 K. Helium was first liquefied by Heike Kamerlingh-Onnes in 1909. He received the Physics Prize in 1913 for the production of liquid helium and for his investigations of properties of matter at low temperatures. Lev D. Landau formulated fundamental concepts (e.g. the 'Landau Liquid') concerning many-body effects in condensed matter and applied them to the theory of liquid helium, explaining specific phenomena occurring in ^{4}He such as superfluidity (see below), the 'roton' excitations, and certain acoustic phenomena. He was awarded the Physics Prize in 1962.

Several of the experimental techniques used for the production and study of low temperature phenomena were developed by Pyotr L. Kapitsa in the 1920s and 30s. He studied many aspects of liquid ^{4}He and showed that it was superfluid (i.e. flowing without friction) below 2.2 K. The superfluid state was later understood to be a manifestation of macroscopic quantum coherence in a Bose–Einstein type of condensate (theoretically predicted in 1920) with many features in common with the superconducting state for electrons in certain conductors. Kapitsa received one-half of the Physics Prize for 1978.

In liquid ^{3}He, additional, unique phenomena show up because each He nucleus has a non-zero spin in contrast to those of ^{4}He. Thus, it is a fermion type of particle, and should not be able to participate in Bose–Einstein condensation, which works only for bosons. However, like in superconductivity (see below) pairs of spin-half particles can form 'quasi-bosons' that can condense into a superfluid phase. Superfluidity in ^{3}He, whose transition temperature is reduced by a factor of a thousand compared to that of liquid ^{4}He, was discovered by David M. Lee, Douglas D. Osheroff and Robert C. Richardson, who received the Physics Prize in 1996. They observed three different superfluid phases, showing complex vortex structures and interesting quantum behavior.

Electrons in condensed matter can be localized to their respective atoms as in insulators, or they can be free to move between atomic sites, as in conductors and semiconductors. In the beginning of the 20th century, it was known that metals emitted electrons when heated to high temperatures, but it was not clear whether this was due only to thermal excitation of the electrons or if chemical interactions with the surrounding gas were also involved. Through experiments carried out in high vacuum, Owen W. Richardson could finally establish that electron emission is a purely thermionic effect and a law based on the velocity distribution of electrons in the metal could be formulated. For this, Richardson received the Physics Prize in 1928 (awarded in 1929.)

The electronic structure determines the electric, magnetic, and optical properties of solids and is also of major importance for their mechanical and thermal behavior. It has been one of the major tasks of physicists in the 20th century to measure the states and dynamics of electrons and model their behavior so as to understand how they organize themselves in various types of solids. It is natural that the most unexpected and extreme manifestations of electron behavior have attracted the strongest interest in the community of solid state physicists. This is also reflected in the Nobel Prize in Physics: several prizes have been awarded for discoveries connected with superconductivity and for some of the very specific effects displayed in certain semiconducting materials.

Superconductivity was discovered as early as 1911 by Kamerlingh-Onnes, who noticed that the electrical resistivity of mercury dropped to less than one billionth of its ordinary value when it was cooled well below a transition temperature T_c, which is about 4 K. As mentioned earlier, he received the Physics Prize in 1913. However, it would take a very long period of time before it was understood why electrons could flow without resistance in certain conductors at low temperature. But in the beginning of the 1960s Leon N. Cooper, John Bardeen and J. Robert Schrieffer formulated a theory based on the idea that pairs of electrons (with opposite spins and directions of motion) can lower their energy by an amount E_g by sharing exactly the same deformation of the crystalline lattice as they move. Such 'Cooper pairs' act as bosonic particles. This allows them to move as a coherent macroscopic fluid, undisturbed as long as the thermal excitations (of energy kT) are lower in energy than the energy E_g gained by the pair formation. The so-called BCS-theory was rewarded with the Physics Prize in 1972.

This breakthrough in the understanding of the quantum mechanical basis led to further progress in superconducting circuits and components: Brian D. Josephson analysed the transfer of superconducting carriers between two superconducting metals, separated by a very thin layer of normal-conducting material. He found that the quantum phase, which determines the transport properties, is an oscillating function of the voltage applied over this kind of junction. The Josephson effect has important applications in precision measurements, since it establishes a relation between voltage and frequency scales. Josephson received one-half of the Physics Prize for 1973. Ivar Giaever, who invented and studied the detailed properties of the 'tunnel junction', an electronic component based on superconductivity, shared the second half with Leo Esaki for work on tunneling phenomena in semiconductors (see below).

Although a considerable number of new superconducting alloys and compounds were discovered over the first 75 years that followed Kamerlingh-Onnes' discovery, it seemed as if superconductivity would forever remain a typical low temperature phenomenon, with the limit for transition temperatures slightly above 20 K. It therefore came as a total surprise when J. Georg Bednorz and K. Alexander Müller showed that a lanthanum-copper oxide could be made superconducting up to 35 K by doping it with small amounts of barium. Soon thereafter, other laboratories reported that cuprates of similar structure were superconducting up to about 100 K. This discovery of 'high temperature superconductors' triggered one of the greatest efforts in modern physics: to understand the basic mechanism for superconductivity in these extraordinary materials. Bednorz and Müller shared the Physics Prize in 1987.

Electron motion in the normal conducting state of metals has been modeled theoretically with increasing degree of sophistication ever since the advent of quantum mechanics. One of the early major steps was the introduction of the Bloch wave concept, named after Felix Bloch (half of the Physics Prize for magnetic resonance in 1952). Another important concept, 'the electron fluid' in conductors, was introduced by Lev Landau (see liquid He). Philip W. Anderson made several important contributions to the theory of electronic structures in metallic systems, in particular concerning the effects of inhomogeneities in alloys and magnetic impurity atoms in metals. Nevill F. Mott worked on the general conditions for electron conductivity in solids and formulated rules for the point at which an insulator becomes a conductor (the Mott transition) when composition or external parameters are changed. Anderson and Mott shared the 1977 Physics Prize with John H. Van Vleck for their theoretical investigations of the electronic structure of magnetic and disordered systems.

An early Physics Prize (1920) was given to Charles E. Guillaume for his discovery that the thermal expansion of certain nickel steels, so-called 'invar' alloys, was practically zero. This prize was mainly motivated by the importance of these alloys for precision measurements in physics and geodesy, in particular when referring to the standard meter in Paris. The invar alloys have been extensively used in all kinds of high-precision mechanical devices, watches, etc. The theoretical background for this temperature independence has been explained only recently. Also very recently (1998), Walter Kohn was recognized by a Nobel Prize in Chemistry for his methods of treating quantum exchange correlations, by which important limitations for the predictive power of electronic structure calculations, in solids as well as molecules, have been overcome.

In semiconductors, electron mobility is strongly reduced because there are forbidden regions for the energy of the electrons that take part in conduction, the 'energy gaps'. It was only after the basic roles of doping of ultra-pure silicon (and later other semiconducting materials) with chosen electron-donating or electron-accepting agents were understood, that semiconductors could be used as components in electronic engineering. William B. Shockley, John Bardeen (see also BCS-theory) and Walter H. Brattain carried out fundamental investigations of semiconductors and developed the first transistor. This was the beginning of the era of 'solid state electronics'. They shared the Physics Prize in 1956.

Later, Leo Esaki developed the tunnel diode, an electronic component that has a negative differential resistance, a technically interesting property. It is composed of two heavily *n* and *p* doped semiconductors, that have an excess of electrons on one side of the junction and a deficit on the other. The tunneling effect occurs at bias voltages larger than the gap in the semi-conductors. He shared the Physics Prize for 1973 with Brian D. Josephson.

With modern techniques it is possible to build up well-defined, thin-layered structures of different semiconducting materials, in direct contact with each other. With such 'heterostructures' one is not limited to the band-gaps provided by natural semi-conducting materials like silicon and germanium. Herbert Kroemer analysed theoretically the mobility of electrons and holes in heterostructure junctions. His propositions led to the build up of transistors with much improved characteristics, later called HEMTs (high electron mobility transistors), which are very important in today's high-speed electronics. Kroemer suggested also, at about the same time as Zhores I. Alferov, the use of double heterostructures to provide conditions for laser action. Alferov later built the first working pulsed semiconductor laser in 1970. This marked the beginning of the era of modern optoelectronic devices now used in laser diodes, CD-players, bar code readers and fiber optics communication. Alferov and Kroemer recently shared one-half of the Physics Prize for the year 2000. The other half went to Jack S. Kilby, co-inventor of the integrated circuit (see the next section on Physics and Technology).

By applying proper electrode voltages to such systems one can form 'inversion layers', where charge carriers move essentially only in two dimensions. Such layers have turned out to have some quite unexpected and interesting properties. In 1982, Klaus von Klitzing discovered the quantized Hall effect. When a strong magnetic field is applied perpendicularly to the plane of a quasi two-dimensional layer, the quantum conditions are such that an increase of magnetic field does not give rise to a linear increase of voltage on the edges of the sample, but a step-wise one. Between these steps, the Hall resistance is h/ie^2, where i's are integers corresponding to the quantized electron orbits in the plane. Since this provides a possibility to measure the ratio between two fundamental constants very exactly, it has important consequences for measurement technology. von Klitzing received the Physics Prize in 1985.

A further surprise came shortly afterwards when Daniel C. Tsui and Horst L. Störmer made refined studies of the quantum Hall effect using inversion layers in materials of ultra-high purity. Plateaus appeared in the Hall effect not only for magnetic fields corresponding to the filling of orbits with one, two, three, etc, electron charges, but also for fields corresponding to fractional charges! This could be understood only in terms of a new kind of quantum fluid, where the motion of independent electrons of charge e is replaced by excitations in a multi–particle system which behave (in a strong magnetic field) as if charges of $e/3$, $e/5$, etc. were involved. Robert B. Laughlin developed the theory that describes this new state of matter and shared the 1998 Physics Prize with Tsui and Störmer.

Sometimes, discoveries made in one field of physics turn out to have important applications in quite different areas. One example, of relevance for solid state physics, is the observation by Rudolf L. Mössbauer in the late 50s, that nuclei in 'absorber' atoms can be resonantly excited by γ rays from suitably chosen 'emitter' atoms, if the atoms in both cases are bound in such a way that recoils are eliminated. The quantized energies of the nuclei in the internal electric and magnetic fields of the solid can be measured since they correspond to different positions of the resonances, which are extremely sharp. This turned out to be important for the determination of electronic and magnetic structure of many substances and Mössbauer received half the Physics Prize in 1961 and R. Hofstadter (see page 43) the other half.

5. Physics and Technology

Many of the discoveries and theories mentioned so far in this survey have had an impact on the development of technical devices; by opening completely new fields of physics or by providing ideas upon which such devices can be built. Conspicuous examples are the works of Shockley, Bardeen, and Brattain which led to the transistors and started a revolution in electronics and the basic research by Townes, Basov, and Prokhorov which led to the development of masers and lasers. It could also be mentioned that particle accelerators are now important tools in several areas of materials science and in medicine. Other works honored by Nobel Prizes have had a more direct technical motivation, or have turned out to be of particular importance for the construction of devices for the development of communication and information.

An early Physics Prize (1912) was given to Nils Gustaf Dalén for his invention of an automatic 'sun-valve', extensively used for lighting beacons and light buoys. It was based on the difference in radiation of heat from reflecting and black bodies: one out of three parallel bars in his device was blackened, which gave rise to a difference in heat absorption and length expansion of the bars during sunshine hours. This effect was used to automatically switch off the gas supply in daytime, eliminating much of the need for maintenance at sea.

Optical instrumentation and techniques have been the topics for prizes at several occasions. Around the turn of the century, Gabriel Lippmann developed a method for colour photograhy using interference of light. A mirror was placed in contact with the emulsion of a photographic plate in such a way that when it was illuminated, reflection in the mirror gave rise to standing waves in the emulsion. Developing resulted in a stratification of the grains of silver and when such a plate was looked at in a mirror, the picture was reproduced in its natural colours. The Physics Prize in 1908 was awarded to Lippmann. Unfortunately, Lippmann's method requires very long exposure times. It has later been superseded by other techniques for photography but has found new applications in high-quality holograms.

In optical microscopy it was shown by Frits Zernike that even very weakly absorbing (virtually transparent) objects can be made visible if they consist of regions with different refractive indices. In Zernike's 'phase-contrast microscope' it is possible to distinguish patches of light that have undergone different phase changes caused by this kind of inhomogeneity. This microscope has been of particular importance for observing details in biological samples. Zernike received the Physics Prize in 1953. In the

1940s, Dennis Gabor laid down the principles of holography. He predicted that if an incident beam of light is allowed to interfere with radiation reflected from a two-dimensional array of points in space, it would be possible to reproduce a three-dimensional picture of an object. However, the realization of this idea had to await the invention of lasers, which could provide the coherent light necessary for such interference phenomena to be observed. Gabor was awarded the Physics Prize in 1971.

Electron microscopy has had an enormous impact on many fields of natural sciences. Soon after the wave nature of electrons was clarified by C. J. Davisson and G. P. Thomson, it was realized that the short wavelengths of high energy electrons would make possible a much increased magnification and resolution as compared to optical microscopes. Ernst Ruska made fundamental studies in electron optics and designed the first working electron microscope early in the 1930s. However, it would take more than 50 years before this was recognized by a Nobel Prize.

Ruska obtained half of the Physics Prize for 1986, while the other half was shared between Gerd Binnig and Heinrich Rohrer, who had developed a completely different way to obtain pictures with extremely high resolution. Their method is applicable to surfaces of solids and is based on the tunneling of electrons from very thin metallic tips to atoms on the surface when the tip is moved at very close distance to it (about 1 nm). By keeping the tunneling current constant a moving tip can be made to follow the topography of the surface, and pictures are obtained by scanning over the area of interest. By this method, single atoms on surfaces can be visualized.

Radio communication is one of the great technical achievements of the 20th century. Guglielmo Marconi experimented in the 1890s with the newly discovered Hertzian waves. He was the first one to connect one of the terminals of the oscillator to the ground and the other one to a high vertical wire, the 'antenna', with a similar arrangement at the receiving station. While Hertz' original experiments were made within a laboratory, Marconi could extend signal transmission to distances of several kilometers. Further improvement was made by Carl Ferdinand Braun (also father of the 'Braunian tube', an early cathode ray oscilloscope), who introduced resonant circuits in the Hertzian oscillators. The tunability and the possibility to produce relatively undamped outgoing oscillations greatly increased the transmission range, and in 1901 Marconi succeeded in establishing radio connection across the Atlantic. Marconi and Braun shared the 1909 Nobel Prize in Physics.

At this stage, it was not understood how radio waves could reach distant places (practically 'on the other side of the earth'), keeping in mind that they were known to be of the same nature as light, which propagates in straight lines in free space. Sir Edward V. Appleton finally proved experimentally that an earlier suggestion by Heaviside and Kennelly, that radio waves were reflected between different layers with different conductance in the atmosphere, was the correct explanation. Appleton measured the interference of the direct and reflected waves at various wavelengths and could determine the height of Heaviside's layer; in addition he found another one at a higher level which still bears his name. Appleton received the Physics Prize in 1947.

Progress in nuclear and particle physics has always been strongly dependent on advanced technology (and sometimes a driving force behind it). This was already illustrated in connection with the works of Cockcroft and Walton and of Lawrence, who developed linear electrostatic accelerators and cyclotrons, respectively. Detection of high energy particles is also a technological challenge, the success of which has been recognized by several Nobel Prizes.

The Physics Prize in 1958 was jointly awarded to Pavel A. Cherenkov, Il'ja M. Frank and Igor Y. Tamm for their discovery and interpretation of the Cherenkov effect. This is the emission of light, within a cone of specific opening angle around the path of a charged particle, when its velocity exceeds the velocity of light in the medium in which it moves. Since this cone angle can be used to determine the velocity of the particle, the work by these three physicists soon became the basis for fruitful detector developments.

The visualization of the paths of particles taking part in reactions is necessary for the correct interpretation of events occurring at high energies. Early experiments at relatively low energies used the tracks left in photographic emulsions. Charles T. R. Wilson developed a chamber in which particles were made visible by the fact that they leave tracks of ionized gas behind them. In the Wilson chamber the gas is made to expand suddenly, which lowers the temperature and leads to condensation of vapour around the ionized spots; these drops are then photographed in strong light. Wilson received half of the Physics Prize in 1927, the other half was awarded to Arthur H. Compton.

A further step in the same direction came much later when Donald A. Glaser invented the 'bubble chamber'. In the 1950s accelerators had reached energies of 20–30 GeV and earlier methods were inadequate; for the Wilson chamber the path lengths in the gas would have been excessive. The atomic nuclei in a bubble chamber (usually containing liquid hydrogen) are used as targets, and the tracks of produced particles can be followed.

At the temperature of operation the liquid is superheated and any discontinuity, like an ionized region, immediately leads to the formation of small bubbles. Essential improvements were made by Luis W. Alvarez, in particular concerning recording techniques and data analysis. His work contributed to a fast extension of the number of known elementary particles then known, in particular the so-called 'resonances' (which were later understood as excited states of systems composed of quarks and gluons). Glaser received the Physics Prize in 1960 and Alvarez in 1968.

Bubble chambers were, up to the end of the 80s, the work horses of all high energy physics laboratories but have later been superseded by electronic detection systems. The latest step in detector development recognized by a Nobel Prize (in 1992) is the work of Georges Charpak. He studied in detail the ionization processes in gases and invented the 'wire chamber', a gas-filled detector where densely spaced wires pick up electric signals near the points of ionization, by which the paths of particles can be followed. The wire chamber and its followers, the time projection chamber and several large wire chamber/scintillator/Cherenkov detector arrangements, combined into complex systems, has made possible the selective search for extremely rare events (like heavy quark production), which are hidden in strong backgrounds of other signals.

The first Nobel Prize (year 2000) in the new millennium was awarded in half to Jack S. Kilby for achievements that laid the foundations for the present information technology. In 1958, he fabricated the first integrated circuit where all electronic components are built on one single block of semiconducting material, later called 'chip'. This opened the way for miniaturization and mass production of electronic circuits. In combination with the development of components based on heterostructures described in an earlier section (for which Alferov and Kroemer shared the other half of the Prize), this has led to the 'IT-revolution' that has reshaped so much our present society.

6. Further Remarks

In reading the present survey, it should be kept in mind that the number of Nobel awards is limited (according to the present rules, at most 3 persons can share a Nobel Prize each year). So far, 163 laureates have received Nobel Prizes for achievements in physics. Often, during the selection process, committees have had to leave out several other important, 'near Nobel-worthy' contributions. For obvious reasons, it has not been possible to mention any of these other names and contributions in this survey. Still, the very fact that a relatively coherent account of the development of physics can be formulated, hinging as here on the ideas and experiments made by Nobel Laureates, can be taken as a testimony that most of the essential features in this fascinating journey towards an understanding of the world we inhabit have been covered by the Nobel Prizes in Physics.

Physics 1901

Wilhelm Conrad Röntgen (1845–1923)

"in recognition of the extraordinary services he has rendered by the discovery of the remarkable rays subsequently named after him"

Physics 1903

Marie Curie (1867–1934)

"in recognition of the extraordinary services rendered on the radiation phenomena discovered by Professor Henri Becquerel"

Physics 1918

Max Karl Ernst Ludwig Planck (1858–1947)

"in recognition of the services he rendered to the advancement of Physics by his discovery of energy quanta"

Physics 1921

Albert Einstein (1879–1955)

"for his services to Theoretical Physics, and especially for his discovery of the law of the photoelectric effect"

Physics 1922

Niels Henrik David Bohr (1885–1962)

"for his services in the investigation of the structure of atoms and of the radiation emanating from them"

Physics 1949

Hideki Yukawa (1907–1981)

"for his prediction of the existence of mesons on the basis of theoretical work on nuclear forces"

Physics 1965

Richard P. Feynman (1918–1988)

"for fundamental work in quantum electrodynamics, with deep-ploughing consequences for the physics of elementary particles"

CHEMISTRY

"...one part to the person who shall have made the most important chemical discovery or improvement..."

The Nobel Prize in Chemistry: The Development of Modern Chemistry

Bo G. Malmström[*] *and Bertil Andersson*[**]

1. Introduction

1.1. *Chemistry at the Borders to Physics and Biology*

The turn of the century 1900 was also a turning point in the history of chemistry. Consequently, a survey of the Nobel Prizes in Chemistry during this century will provide an analysis of important trends in the development of this branch of the Natural Sciences, and this is the aim of the present essay. Chemistry has a position in the center of the sciences, bordering onto physics, which provides its theoretical foundation, on one side, and onto biology on the other, living organisms being the most complex of all chemical systems. Thus, the fact that chemistry flourished during the beginning of the 20th century is intimately connected with fundamental developments in physics.

In 1897 Sir Joseph John Thomson (1856–1940) of Cambridge announced his discovery of the electron, for which he was awarded the Nobel Prize for Physics in 1906. He found that these negatively charged 'corpuscles', as he called them, have a mass 1000 times smaller than the hydrogen atom. Thomson's discovery had, of course, important implications for chemistry, as it showed that the atom is not an indivisible building block of chemical compounds, but it took a number of years before this

[*]Was Professor of Biochemistry at Göteborg University, Sweden. Deceased in 2000.
[**]Professor of Biochemistry and Principal of Linköping University, Sweden. Andersson has contributed with an update of the 1999 and 2000 Prizes.

led to developments of direct relevance to chemistry. In 1911 Ernest Rutherford (1871–1937), who had worked in Thomson's laboratory in the 1890s, formulated an atomic model, according to which the positively charged atomic nucleus carries most of the mass of the atom but occupies a very small part of its volume.

This is instead created by a cloud of electrons circling around the nucleus. Rutherford received the Nobel Prize for Chemistry already in 1908 for his work on radioactivity (see Section 2).

It was soon realized that in Rutherford's atomic model the stability of atoms was at variance with the laws of classical physics, since the electrons would lose energy in the form of electromagnetic radiation and eventually fall into the nucleus. Niels Bohr (1885–1962) from Copenhagen understood that an important clue to the solution of this problem could be found in the distinct lines observed in the spectra of atoms, the regularities of which had been discovered in 1890 by the physics professor Johannes (Janne) Rydberg (1854–1919) at Lund University. Consequently, Bohr formulated in 1913 an alternative atomic model, in which only certain circular orbits of the electrons are allowed. In this model light is emitted (or absorbed), when an electron makes a transition from one orbit to another. Bohr received the Nobel Prize for Physics in 1922 for his work on the structure of atoms.

Another step in the application of the electronic structure of atoms to chemistry was taken in 1916, when Gilbert Newton Lewis (1875–1946) suggested that strong (covalent) bonds between atoms involve a sharing of two electrons between these atoms (electron-pair bond). Lewis also contributed fundamental work in chemical thermodynamics, and his brilliant textbook, *Thermodynamics* (1923), written together with Merle Randall (1880–1950), is counted as one of the masterworks in the chemical literature. Much to the surprise of the chemical community, Lewis never received a Nobel Prize.

Even if the contributions just described were made a decade or more after Thomson's discovery, much important work in the borderland between physics and chemistry was published in the 1890s, and this was naturally given a strong consideration by the first Nobel Committee for Chemistry (see Section 2). In fact, three of the Laureates during the first decade, Jacobus Henricus van't Hoff (1852–1911), Svante Arrhenius (1859–1927) and Wilhelm Ostwald (1853–1932), are generally regarded as the founders of a new branch of chemistry, physical chemistry. Fundamental work had, however, also been done in more traditional chemical fields, particularly in organic chemistry and in the chemistry of natural products, which is clearly reflected in the early prizes. The Nobel Committee, in addition, showed great openness and foresight by recognizing the other border, that towards biology, already in 1907 with the prize to Eduard Buchner (1860–1917) "for his biochemical researches and his discovery of cell-free fermentation."

1.2. *The Mechanics of the Work in the Nobel Committee for Chemistry*

According to the statutes of the Nobel Foundation, the Nobel Committees should have five members, but the Committee for Chemistry has in recent decades chosen to widen its expertise by adding a number of adjunct members (five in 1998) with the same voting rights as the regular members. Until recently there was no limit other than age on how many times regular members could be re-elected for 3-year terms, so that some members sat on the Committee for a very long period. For example, Professor Arne Westgren (1889–1975) of Stockholm, who was secretary of the Nobel Committees for Physics and for Chemistry 1926–1943, was also Chairman of the Committee for Chemistry 1944–1965. Present rules, however, only allow two re-elections, so that a member's maximum total time on the Committee will be nine years.

Only persons who have been properly nominated before 31 January can be considered for the Nobel Prize in a given year. Consequently, the Nobel Committee starts its work by sending out invitations to nominate in the autumn of the preceding year. Recipients of these invitations, for both Physics and Chemistry, are: 1) Swedish and foreign members of the Royal Swedish Academy of Sciences; 2) members of the Nobel Committees for Physics and for Chemistry; 3) Nobel Laureates in Physics and Chemistry; 4) professors in Physics and Chemistry in Scandinavian universities and at Karolinska Institutet; 5) professors in these subjects in a number of universities outside Scandinavia, selected on a rotation basis by the Academy of Sciences; and 6) other scientists that the Academy chooses to invite.

In the initial years of the Nobel Prize, about 300 invitations to nominate for the Nobel Prize for Chemistry were sent out, but this number has increased over the years and it was as high as 2,650 in 1998. The number of nominations received has also increased dramatically from 20–40 during the first decade to 400–500 in the 1990s. The number of candidates is usually smaller than the number of nominations, since many candidates receive more than one nomination. During the first few years only about 10 scientists were nominated, but in recent years this number has been in the range of 250–350.

The invitations to nominate are personal, and it is stressed that nominations should not be discussed with the candidate or with colleagues. This is unfortunately not always respected as is obvious from the fact that many identically worded nominations are some years received from the same university. For this reason the Committee does not put much weight on the number of nominations a given candidate receives, unless clearly independent nominations come from different universities in different countries. This attitude was not taken in earlier years however, as is evident from the following statement made by Committee Chairman Arne Westgren, in a survey over the first 60 years of the Nobel Prize for Chemistry [1]: "In fact, if a scientist is proposed by a large number of sponsors in the preliminary international voting, he is normally selected by the Academy."

Often the same candidate receives nominations both for chemistry and for physics or for chemistry and for medicine. This problem was met already in 1903, when Arrhenius had been nominated both for the Prize for Chemistry and that for Physics, and in its deliberations the Committee for Chemistry suggested that he should be awarded half of each Prize, but this idea was rejected by the Committee for Physics. Because of such borderline problems, the Committee for Chemistry nowadays has joint meetings with those for Physics and for Physiology or Medicine. However, as pronounced by Westgren [1]: "It is now generally recognized that the important thing is to decide whether work which can with equal justice be reckoned as chemistry and physics or chemistry and medicine, is in fact worthy of a Nobel Prize." For example, Peter Mitchell (1920–1992), who received the 1978 Nobel Prize for Chemistry, could with equal justice have been awarded the Prize for Physiology or Medicine.

Nobel's will laid down that the prize should be awarded for work done during the preceding year, but in the statutes governing the committee work this has been interpreted to mean the most recent results, or for older work provided its significance has only recently been demonstrated. It was undoubtedly this rule that excluded Stanislao Cannizzaro (1826–1910) from receiving one of the first Nobel Prizes, since his work on drawing up a reliable table of atomic weights, helping to establish the periodic system, was done in the middle of the 19th century. A more recent example is Henry Eyring (1901–1982), whose brilliant theory for the rates of chemical reactions, published in 1935, was apparently not understood by members of the Nobel Committee until much later. As a compensation the Royal Swedish Academy of Sciences gave him, in 1977, its highest honor, other than the Nobel Prize, the Berzelius Medal in gold.

2. The First Decade of Nobel Prizes for Chemistry

So much fundamental work in chemistry had been carried out during the last two decades of the 19th century that, as stated by Westgren [1], "During the first few years the Academy was chiefly faced with merely deciding the order in which these scientists should be awarded the prize." For the first prize in 1901 the Academy had to consider 20 nominations, but no less than 11 of these named van't Hoff, who was also chosen by the Committee for Chemistry. van't Hoff had already during his thesis work in Utrecht in 1874 published his suggestion that the carbon atom has its four valences directed towards the corners of a regular tetrahedron, a concept which is the very foundation of modern organic chemistry. The Nobel Prize was, however, awarded for his later work on chemical kinetics and equilibria and on the osmotic pressure in solution, published in 1884 and 1886, when he held a professorship in Amsterdam. When he received the prize he had, however, left this for a position at *Akademie der Wissenschaften* in Berlin in 1896.

In his 1886 work van't Hoff showed that most dissolved chemical compounds give an osmotic pressure equal to the gas pressure they would have exerted in the absence of the solvent. An apparent exception was aqueous solutions of electrolytes (acids, bases and their salts), but in the following year Arrhenius showed that this anomaly could be explained, if it is assumed that electrolytes in water dissociate into ions. Arrhenius had already presented the rudiments of his dissociation theory in his doctoral thesis, which was defended in Uppsala in 1884 and was not entirely well received by the faculty. It was, however, strongly supported by Ostwald in Riga, who, in fact, travelled to Uppsala to initiate a collaboration with Arrhenius. In 1886–1890 Arrhenius did work with Ostwald, first in Riga and then in Leipzig, and also with van't Hoff in Berlin. When Arrhenius was awarded the Nobel Prize for Chemistry in 1903, he was since 1895 professor of physics in Stockholm, and he was also nominated for the Prize for Physics (see Section 1).

The award of the Nobel Prize for Chemistry in 1909 to Ostwald was chiefly in recognition of his work on catalysis and the rates of chemical reactions. Ostwald had in his investigations, following up observations in his thesis in 1878, shown that the rate of acid–catalyzed reactions is proportional to the square of the strength of the acid, as measured by titration with base. His work offered support not only to Arrhenius' theory of dissociation but also to van't Hoff's theory for osmotic pressure. Ostwald was founder and editor of *Zeitschrift für Physikalische Chemie*, the publication of which is generally regarded as the birth of this new branch of chemistry.

Three of the Nobel Prizes for Chemistry during the first decade were awarded for pioneering work in organic chemistry. In 1902 Emil Fischer (1852–1919), then in Berlin, was given the prize for "his work on sugar and purine syntheses." Fischer's work is an example of the growing interest from organic chemists in biologically important substances, thus laying the foundation for the development of biochemistry, and at the time of the award Fischer mainly devoted himself to the study of proteins. Another major influence from organic chemistry was the development of chemical industry, and a chief contributor here was Fischer's teacher, Adolf von Baeyer (1835–1917) in Munich, who was awarded the prize in 1905 "in recognition of his services in the advancement of organic chemistry and the chemical industry, … ." His contributions include, in particular, structure determination of organic dyes (indigo, eosin) and the study of aromatic compounds (terpenes). The third Laureate working in organic chemistry was Otto Wallach (1847–1931) in Göttingen, who, like von Baeyer, contributed to alicyclic chemistry, studying not only terpenes but also camphor and other components of ethereal oils. At the award ceremony in 1910 the importance of his discoveries for chemical industry was emphasized.

Two of the early prizes were given for the discovery of new chemical elements. Sir William Ramsay (1852–1916) from London received the 1904 Nobel Prize for Chemistry for his discovery of a number of noble gases, a new group of chemically unreactive elements. The first one isolated was argon ('the inactive one'), which Ramsay discovered in 1894, in collaboration with Lord Rayleigh [John William Strutt Rayleigh (1842–1919)] of the Royal Institution, who was awarded the Prize for Physics in the same year, his investigations of the density of air and other gases forming the basis for this discovery. The following year Ramsay found helium, observed earlier only in the solar spectrum (hence its name), in emanations from radium, thus anticipating later prizes for nuclear chemistry (see below). Later (1898) he also discovered, by fractional distillation of liquid air, neon ('the new one'), krypton ('the hidden one') and xenon ('the strange one'). The isolation of another element, fluorine, by Henri Moissan (1852–1907) in Paris was honored with the 1906 Nobel Prize. In attempts to prepare artificial diamonds Moissan had also developed an electric furnace, and this was specifically mentioned in the prize citation, perhaps a reflection of the stipulation in Nobel's will that the Prize for Chemistry can be given "for the most important discovery or improvement."

Ernest Rutherford [Lord Rutherford since 1931], professor of physics in Manchester, was awarded the Nobel Prize for Chemistry in 1908 for his investigations of the chemistry of radioactive substances. The discovery of radioactivity had already been recognized with the Nobel Prize for Physics in 1903, but what Rutherford established was the transformation of one element into another, earlier the alchemist's dream. In his studies of uranium disintegration he found two types of radiation, named α- and β-rays, and by their deviation in electric and magnetic fields he could show that α-rays consist of positively charged particles. His demonstration that these particles are helium nuclei came in the same year as he received the Nobel Prize. Even if the importance of Rutherford's work for chemistry is obvious, he naturally had also received many nominations for the Nobel Prize for Physics (see Section 1).

In 1897 Eduard Buchner, at the time professor in Tübingen, published results demonstrating that the fermentation of sugar to alcohol and carbon dioxide can take place in the absence of yeast cells. Earlier it had generally been considered that living cells possess a 'vital force', which makes the life processes possible, even if a few prominent chemists, foremost Jöns Jacob Berzelius (1779–1848) and Justus von Liebig (1803–1873), had advocated a chemical basis for life. The vitalistic outlook had been fiercely defended by Louis Pasteur (1822–1895), who maintained that alcoholic fermentation can only occur in the presence of living yeast cells. Buchner's experiments showed unequivocally that fermentation is a catalytic process caused by the action of enzymes, as had been suggested by Berzelius for all life processes, and Buchner called his extract zymase ('enzymes in yeast'). Because of Buchner's experiment, 1897 is generally regarded as the birth date for biochemistry proper. Buchner was awarded the Nobel Prize for Chemistry in 1907, when he was professor at the agricultural college in Berlin. This confirmed the prediction of his former teacher, Adolf von Baeyer: "This will make him famous, in spite of the fact that he lacks talent as a chemist."

3. The Nobel Prizes for Chemistry 1911–2000

A survey of the Nobel Prizes for Chemistry awarded during the 20th century, reveals that the development of this field includes breakthroughs in all of its branches, with a certain dominance for progress in physical chemistry and its subcategories (chemical thermodynamics and chemical change), in chemical structure, in several areas of organic chemistry as well as in biochemistry. Of course, the borders between different areas are diffuse, therefore many Laureates will be mentioned in more than one place.

3.1. *General and Physical Chemistry*

The Nobel Prize for Chemistry in 1914 was awarded to Theodore William Richards (1868–1928) of Harvard University for "his accurate determinations of the atomic weight of a large number of chemical elements." Most atomic weights in Cannizzaros table (see Section 1.2) had already been determined in the 19th century, particularly by the Belgian chemist Jean Servais Stas (1813–1891), but Richards showed that many of them were in error, mainly because Stas had worked with very concentrated solutions, leading to co-precipitation. In 1913 Richards had discovered that the atomic weight of natural lead and of that formed in radioactive decay of uranium minerals differ. This pointed to the existence of isotopes, i.e. atoms of the same element with different atomic weights, which was accurately demonstrated by Francis William Aston (1877–1945) at Cambridge University, with the aid of an instrument developed by him, the mass spectrograph. Aston also showed that the atomic weights of pure isotopes are integral numbers, with the exception of hydrogen, the atomic weight of which is 1.008. For his achievements Aston received the Nobel Prize for Chemistry in 1922.

One branch of physical chemistry deals with chemical events at the interface of two phases, for example, solid and liquid, and phenomena at such interfaces have important applications all the way from technical to physiological processes. Detailed studies of adsorption on surfaces, were carried out by Irving Langmuir (1881–1957) at the research laboratory of General Electric Company, and when he was awarded the Nobel Prize for Chemistry in 1932, he was the first industrial scientist to receive this distinction.

Two of the Prizes for Chemistry in more recent decades have been given for fundamental work in the application of spectroscopic methods to chemical problems. Spectroscopy had already been recognized with Prizes for Physics in 1952, 1955 and 1961, when Gerhard Herzberg (1904–1999), a physicist at the University of Saskatchewan, received the Nobel Prize for Chemistry in 1971 for his molecular spectroscopy studies "of the electronic structure and geometry of molecules, particularly free radicals." The most used spectroscopic method in chemistry is undoubtedly NMR (nuclear magnetic resonance), and Richard R. Ernst (1933–) at ETH in

Zürich was given the Nobel Prize for Chemistry in 1991 for "the development of the methodology of high resolution nuclear magnetic resonance (NMR) spectroscopy." Ernst's methodology has now made it possible to determine the structure in solution (in contrast to crystals; cf. Section 3.5) of large molecules, such as proteins.

3.2. *Chemical Thermodynamics*

The first Nobel Prize for Chemistry, that to van't Hoff, was in part for work in chemical thermodynamics, and many later contributions in this area have also been recognized with Nobel Prizes. Already in 1920 Walther Hermann Nernst (1864–1941) of Berlin received this award for work in thermochemistry, despite a 16-year opposition to this recognition from Arrhenius [2]. Nernst had shown that it is possible to determine the equilibrium constant for a chemical reaction from thermal data, and in so doing he formulated what he himself called the third law of thermodynamics. This states that the entropy, a thermodynamic quantity, which is a measure of the disorder in the system, approaches zero as the temperature goes towards absolute zero. Van't Hoff had derived the mass action equation in 1886, with the aid of the second law which says, that the entropy increases in all spontaneous processes [this had already been done in 1876 by J. Willard Gibbs (1839–1903) at Yale, who certainly had deserved a Nobel Prize, but his work had been published in an obscure place]. According to the second law, heat of reaction is not an accurate measure of chemical equilibrium, as had been assumed by earlier investigators. But Nernst showed in 1906 that it is possible with the aid of the third law, to derive the necessary parameters from the temperature dependence of thermochemical quantities.

To prove his heat theorem (the third law) Nernst carried out thermochemical measurements at very low temperatures, and such studies were extended in the 1920s by G. N. Lewis (see Section 1.1) in Berkeley. Lewis's new formulation of the third law was confirmed by his student William Francis Giauque (1895–1982), who extended the temperature range experimentally accessible by introducing the method of adiabatic demagnetization in 1933. With this he managed to reach temperatures a few thousandths of a degree above absolute zero and could thereby provide extremely accurate entropy estimates. He also showed that it is possible to determine entropies from spectroscopic data. Giauque was awarded the Nobel Prize for Chemistry in 1949 for his contributions to chemical thermodynamics.

The next Nobel Prize given for work in thermodynamics went to Lars Onsager (1903–1976) of Yale University in 1968 for contributions to the thermodynamics of irreversible processes. Classical thermodynamics deals with systems at equilibrium, in which the chemical reactions are said to be reversible, but many chemical systems, for example, the most complex of all, living organisms, are far from equilibrium and their reactions are said to be irreversible. With the aid of statistical mechanics Onsager developed in 1931 his so-called reciprocal relations, describing the flow of matter and energy in such systems, but the importance of his work was not recognized until the end of the 1940s. A further step forward in the development of non-equilibrium thermodynamics was taken by Ilya Prigogine (1917–) in Brussels, whose theory of dissipative structures was awarded the Nobel Prize for Chemistry in 1977.

3.3. *Chemical Change*

The chief method to get information about the mechanism of chemical reactions is chemical kinetics, i.e. measurements of the rate of the reaction as a function of reactant concentrations as well as its dependence on temperature, pressure and reaction medium. Important work in this area had been done already in the 1880s by two of the early Laureates, van't Hoff and Arrhenius, who showed that it is not enough for molecules to collide for a reaction to take place. Only molecules with sufficient kinetic energy in the collision do, in fact, react, and Arrhenius derived an equation in 1889 allowing the calculation of this activation energy from the temperature dependence of the reaction rate. With the advent of quantum mechanics in the 1920s (see Section 3.4), Eyring developed his transition-state theory in 1935 and this showed that the activation entropy is also important. Strangely, Eyring never received a Nobel Prize (see Section 1.2).

In 1956 Sir Cyril Norman Hinshelwood (1897–1967) of Oxford and Nikolay Nikolaevich Semenov (1896–1986) from Moscow shared the Nobel Prize for Chemistry "for their researches into the mechanism of chemical reactions." Among Hinshelwood's major contributions his detailed elucidation of the mechanism for the reaction between oxygen and hydrogen can be mentioned, whereas Semenov's award was for his studies of so-called chain reactions.

A limit in investigating reaction rates is set by the speed with which the reaction can be initiated. If this is done by rapid mixing of the reactants, the time limit is about one thousandth of a second (millisecond). In the

1950s Manfred Eigen (1927–) from Göttingen developed chemical relaxation methods that allow measurements in times as short as a thousandth or a millionth of a millisecond (microseconds or nanoseconds). The methods involve disturbing an equilibrium by rapid changes in temperature or pressure and then follow the passage to a new equilibrium. Another way to initiate some reactions rapidly is flash photolysis, i.e. by short light flashes, a method developed by Ronald G. W. Norrish (1897–1978) at Cambridge and George Porter (Lord Porter since 1990) (1920–) in London. Eigen received one-half and Norrish and Porter shared the other half of the Nobel Prize for Chemistry in 1967. The milli- to picosecond time scales gave important information on chemical reactions. However, it was not until it was possible to generate femtosecond laser pulses (10^{-15} s) that it became possible to reveal when chemical bonds are broken and formed. Ahmed Zewail (born 1946 in Egypt) at California Institute of Technology received the Nobel Prize for Chemistry in 1999 for his development of 'femtochemistry' and in particular for being the first to experimentally demonstrate a transition state during a chemical reaction. His experiments relate back to 1889 when Arrhenius (Nobel Prize, 1903) made the important prediction that there must exist intermediates (transition states) in the transformation from reactants to products. Henry Taube (1915–) of Stanford University was awarded the Nobel Prize for Chemistry in 1983 "for his work on the mechanism of electron transfer reactions, especially in metal complexes." Even if Taube's work was on inorganic reactions, electron transfer is important in many catalytic processes used in industry and also in biological systems, for example, in respiration and photosynthesis. The latest prize for work in chemical kinetics was that to Dudley R. Herschbach (1932–) at Harvard University, Yuan T. Lee (1936–) of Berkeley and John C. Polanyi (1929–) from Toronto in 1986. Herschbach and his student Lee introduced the use of fluxes of molecules with well-defined direction and energy, molecular beams. By crossing two such beams they could study details of the reaction between molecules at extremely short times. Another important method to investigate such reaction details is infrared chemiluminescence, introduced by Polanyi. The emission of infrared radiation from the reaction products gives information on the energy distribution in the molecules.

3.4. *Theoretical Chemistry and Chemical Bonding*

Quantum mechanics, developed in the 1920s, offered a tool towards a more basic understanding of chemical bonds. In 1927 Walter Heitler

(1884–1981) and Fritz London (1900–1954) showed that it is possible to solve exactly the relevant equations for the hydrogen molecule ion, i.e. two hydrogen nuclei sharing a single electron, and thereby calculate the attractive force between the nuclei. For molecules containing more than three elementary particles, even the hydrogen molecule with Lewis's two-electron bond (see Section 1.1), the equation can, however, not be solved exactly, so one has to resort to approximate methods. A pioneer in developing such methods was Linus Pauling (1901–1994) at the California Institute of Technology, who was awarded the Nobel Prize for Chemistry in 1954 "for his research into the nature of the chemical bond … ." Pauling's valence-bond (VB) method is rigorously described in his 1935 book *Introduction to Quantum Mechanics* (written together with E. Bright Wilson, Jr., at Harvard). A few years later (1939) he published an extensive non-mathematical treatment in *The Nature of the Chemical Bond*, a book which is one of the most read and influential in the entire history of chemistry. Pauling was not only a theoretician, but he also carried out extensive investigations of chemical structure by X-ray diffraction (see Section 3.5). On the basis of results with small peptides, which are building blocks of proteins, he suggested the α-helix as an important structural element. Pauling was awarded the Nobel Peace Prize for 1962, and he is the only person to date to have won two unshared Nobel Prizes.

Pauling's VB method cannot give an adequate description of chemical bonding in many complicated molecules, and a more comprehensive treatment, the molecular-orbital (MO) method, was introduced already in 1927 by Robert S. Mulliken (1896–1986) from Chicago and later developed further by him as well as by many other investigators. MO theory considers, in quantum-mechanical terms, the interaction between all atomic nuclei and electrons in a molecule. Mulliken also showed that a combination of MO calculations with experimental (spectroscopic) results provides a powerful tool for describing bonding in large molecules. Mulliken received the Nobel Prize for Chemistry in 1966.

Theoretical chemistry has also contributed significantly to our understanding of chemical reaction mechanisms. In 1981 the Nobel Prize for Chemistry was shared between Kenichi Fukui (1918–1998) in Kyoto and Roald Hoffmann (1937–) of Cornell University "for their theories, developed independently, concerning the course of chemical reactions." Fukui introduced in 1952 the frontier–orbital theory, according to which the occupied MO with the highest energy and the unoccupied one with the lowest energy have a dominant influence on the reactivity of a molecule. Hoffmann formulated in 1965, together with Robert B. Woodward

(1917–1979) (see Section 3.8), rules based on the conservation of orbital symmetry, for the reactivity and stereochemistry in chemical reactions.

Rudolph A. Marcus (1923–) published during ten years, starting in 1956, a series of seminal papers on a comprehensive theory for the rates electron-transfer reactions, the experimental study of which had given Taube a Nobel Prize in 1983 (see Section 3.3). Marcus's theory predicts how the rate varies with the driving force for the reaction, i.e. the difference in energy between reactants and products, and counter to intuition he found that it does not increase continuously, but goes through a maximum, into the Marcus inverted region, which has later been confirmed experimentally. Marcus was awarded the Nobel Prize for Chemistry in 1992.

The latest Nobel Prize for work in theoretical chemistry was given in 1998 to Walter Kohn (1923–) of Santa Barbara and John A. Pople (1925–) of Northwestern University (but a British citizen). The prize to Kohn, a theoretical physicist, was based on his development of density-functional theory, which facilitates detailed calculations both of the geometrical structures of complex molecules and of the energy map of chemical reactions. Pople, a mathematician (but now Professor of Chemistry), was awarded "for his development of computational methods in quantum chemistry." In particular, Pople has designed computer programs based on classical quantum theory as well as on density-functional theory.

3.5. *Chemical Structure*

The most commonly used method to determine the structure of molecules in three dimensions is X-ray crystallography. The diffraction of X-rays was discovered by Max von Laue (1879–1960) in 1912, and this gave him the Nobel Prize for Physics in 1914. Its use for the determination of crystal structure was developed by Sir William Bragg (1862–1942) and his son, Sir Lawrence Bragg (1890–1971), and they shared the Nobel Prize for Physics in 1915. The first Nobel Prize for Chemistry for the use of X-ray diffraction went to Petrus (Peter) Debye (1884–1966), then of Berlin, in 1936. Debye did not study crystals, however, but gases, which give less distinct diffraction patterns. He also employed electron diffraction and the measurement of dipole moments to get structural information. Dipole moments are found in molecules, in which the positive and negative charge is unevenly distributed (polar molecules).

Many Nobel Prizes have been awarded for the determination of the structure of biological macromolecules (proteins and nucleic acids). Proteins are long chains of amino-acids, as shown by Emil Fischer (see Section 2), and the first step in the determination of their structure is to determine the order (sequence) of these building blocks. An ingenious method for this tedious task was developed by Frederick Sanger (1918–) of Cambridge, and he reported the amino-acid sequence for a protein, insulin, in 1955. For this achievement he was awarded the Nobel Prize for Chemistry in 1958. Sanger later (1980) received part of a second Nobel Prize for Chemistry for a method to determine the nucleotide sequence in nucleic acids (see Section 3.12), and he is the only scientist so far who has won two Nobel Prizes for Chemistry.

The first protein crystal structures were reported by Max Perutz (1914–) and Sir John Kendrew (1917–1997) in 1960, and these two investigators shared the Nobel Prize for Chemistry in 1962. Perutz had started studying the oxygen-carrying blood pigment, hemoglobin, with Sir Lawrence Bragg in Cambridge already in 1937, and ten years later he was joined by Kendrew, who looked at crystals of the related muscle pigment, myoglobin. These proteins are both rich in Pauling's α-helix (see Section 3.4), and this made it possible to discern the main features of the structures at the relatively low resolution first used. The same year that Perutz and Kendrew won their prize, the Nobel Prize for Physiology or Medicine went to Francis Crick (1916–), James Watson (1928–) and Maurice Wilkins (1916–) "for their discoveries concerning the molecular structure of nucleic acids … ." Two years later (1964) Dorothy Crowfoot Hodgkin (1910–1994) received the Nobel Prize for Chemistry for determining the crystal structures of penicillin and vitamin B_{12}.

Two later Nobel Prizes for Chemistry in the crystallographic field were given for work on structures of relatively small molecules. William N. Lipscomb (1919–) of Harvard received the prize in 1976 "for his studies on the structures of boranes illuminating problems of chemical bonding." In 1985 Herbert A. Hauptman (1917–) of Buffalo and Jerome Karle (1918–) of Washington, DC, shared the prize for "the development of direct methods for the determination of crystal structures." Their methods are called direct, because they yield the structure directly from the diffraction data collected, and they have been indispensable in the determination of the structures of a large number of natural products.

Crystallographic electron microscopy was developed by Sir Aaron Klug (1926–) in Cambridge, who was awarded the Nobel Prize for Chemistry in 1982. With this technique Klug has investigated the structure of large nucleic acid-protein complexes, such as viruses and chromatin, the carrier of the genes in the cell nucleus. Many of the most important life processes are carried out by proteins associated with biological membranes. This is, for example, true of the two key processes in energy metabolism, respiration and photosynthesis. Attempts to prepare crystals of membrane proteins for structural studies were, however, for many years unsuccessful, but in 1982 Hartmut Michel (1948–), then at the Max-Planck-Institut in Martinsried, managed to crystallize a photosynthetic reaction center after a painstaking series of experiments. He then proceeded to determine the three-dimensional structure of this protein complex in collaboration with Johann Deisenhofer (1943–) and Robert Huber (1937–), and this was published in 1985. Deisenhofer, Huber and Michel shared the Nobel Prize for Chemistry in 1988. Michel has later also crystallized and determined the structure of the terminal enzyme in respiration, and his two structures have allowed detailed studies of electron transfer (cf. Sections 3.3 and 3.4) and its coupling to proton pumping, key features of the chemiosmotic mechanism for which Peter Mitchell had already received the Nobel Prize for Chemistry in 1978 (see Section 3.12). Functional and structural studies on the enzyme ATP synthase, connected to this proton pumping mechanism, was awarded one-half of the Nobel Prize for Chemistry in 1997, shared between Paul D. Boyer and John Walker (see Section 3.12).

3.6. *Inorganic and Nuclear Chemistry*

Much of the progress in inorganic chemistry during the 20th century has been associated with investigations of coordination compounds, i.e. a central metal ion surrounded by a number of coordinating groups, called ligands. In 1893 Alfred Werner (1866–1919) in Zürich presented his coordination theory, and in 1905 he summarized his investigations in this new field in a book (*Neuere Anschauungen auf dem Gebiete der anorganischen Chemie*), which appeared in no less than five editions from 1905–1923. Compounds in which a metal ion binds several other molecules (ligands), for example, ammonia, had earlier been thought to have a linear structure, in accord with a theory advanced by the Swedish chemist Wilhelm Blomstrand (1826–1862) in Lund. Werner showed that such a structure is inconsistent with some experimental facts, and he suggested instead that all the ligand molecules are bound directly to the metal ion. Werner was awarded the Nobel Prize for Chemistry in 1913. Taube's investigations of electron transfer, awarded in 1983 (see Section 3.3), were mainly carried out with coordination compounds, and vitamin B_{12} as well as the proteins hemoglobin and myoglobin, investigated by the Laureates Hodgkin, Perutz and Kendrew (see Section 3.5), also belong to this category.

Another early prize for work in inorganic chemistry was that to Fritz Haber (1868–1934) from Berlin in 1918 "for the synthesis of ammonia from its elements," i.e. from nitrogen and hydrogen. The importance of this synthesis is above all in its industrial application in the form of the Haber–Bosch method, which had been developed by Carl Bosch (1874–1940) as an improvement (cf. Nobel's will) of Haber's original procedure. It allows the manufacture of ammonia on a large scale, and the ammonia can then be used for the production of many different nitrogen-containing chemicals. Bosch shared the Nobel Prize for Chemistry with Friedrich Bergius in 1931 (see Section 3.13).

Much inorganic chemistry in the early 1900s was a consequence of the discovery of radioactivity in 1896, for which Henri Becquerel (1852–1908) from Paris was awarded the Nobel Prize for Physics in 1903, together with Pierre (1859–1906) and Marie Curie (1867–1934). In 1911 Marie Curie received the Nobel Prize for Chemistry for her discovery of the elements radium and polonium and for the isolation of radium and studies of its compounds, and this made her the first investigator to be awarded two Nobel Prizes. The prize in 1921 went to Frederick Soddy (1877–1956) of Oxford for his work on the chemistry of radioactive substances and on the origin of isotopes. In 1934 Frédéric Joliot (1900–1958) and his wife Irène Joliot–Curie (1897–1956), the daughter of the Curies, discovered artificial radioactivity, i.e. new radioactive elements produced by the bombardment of non-radioactive elements with α-particles or neutrons. They were awarded the Nobel Prize for Chemistry in 1935 for "their synthesis of new radioactive elements."

Many elements are mixtures of non-radioactive isotopes (see Section 3.1), and in 1934 Harold Urey (1893–1981) of Columbia University had been given the Nobel Prize for Chemistry for his isolation of heavy hydrogen (deuterium). Urey had also separated uranium isotopes, and his work was an important basis for the investigations by Otto Hahn (1879–1968) from Berlin. In attempts to make transuranium elements, i.e. elements with a higher atomic number than 92 (uranium), by radiating uranium atoms with neutrons, Hahn discovered that one of the products was barium, a lighter element. Lise Meitner (1878–1968), at the time a refugee from Nazism in Sweden, who had earlier worked with Hahn and taken the initiative for the uranium bombardment experiments, provided the explanation, namely, that the uranium atom was cleaved and that barium was one of the products

[3]. Hahn was awarded the Nobel Prize for Chemistry in 1944 "for his discovery of the fission of heavy nuclei," and it can be wondered why Meitner was not included. Hahn's original intention with his experiments was later achieved by Edwin M. McMillan (1907–1991) and Glenn T. Seaborg (1912–1999) of Berkeley, who were given the Nobel Prize for Chemistry in 1951 for "discoveries in the chemistry of transuranium elements."

The use of stable as well as radioactive isotopes have important applications, not only in chemistry, but also in fields as far apart as biology, geology and archeology. In 1943 George de Hevesy (1885–1966) from Stockholm received the Nobel Prize for Chemistry for his work on the use of isotopes as tracers, involving studies in inorganic chemistry and geochemistry as well as on the metabolism in living organisms. The prize in 1960 was given to Willard F. Libby (1908–1980) of the University of California, Los Angeles (UCLA), for his method to determine the age of various objects (of geological or archeological origin) by measurements of the radioactive isotope carbon-14.

3.7. *General Organic Chemistry*

Contributions in organic chemistry have led to more Nobel Prizes for Chemistry than work in any other of the traditional branches of chemistry. Like the first prize in this area, that to Emil Fischer in 1902 (see Section 2), most of them have, however, been awarded for advances in the chemistry of natural products and will be treated separately (Section 3.9). Another large group, preparative organic chemistry, has also been given its own section (Section 3.8), and here only the prizes for more general contributions to organic chemistry will be discussed. In 1969 the Nobel Prize for Chemistry went to Sir Derek H. R. Barton (1918–1998) from London, and Odd Hassel (1897–1991) from Oslo for developing the concept of conformation, i.e. the spatial arrangement of atoms in molecules, which differ only by the orientation of chemical groups by rotation around a single bond. This stereochemical concept rests on the original suggestion by van't Hoff of the tetrahedral arrangement of the four valences of the carbon atom (see Section 2), and most organic molecules exist in two or more stable conformations.

The Nobel Prize for Chemistry in 1975 to Sir John Warcup Cornforth (1917–) of the University of Sussex and Vladimir Prelog (1906–1998) of ETH in Zürich was also based on research in stereochemistry. Not only can a compound have more than one geometric form, but chemical reactions can also have specificity in their stereochemistry, thereby forming a product with a particular three-dimensional arrangement of the atoms. This is especially true of reactions in living organisms, and Cornforth has mainly studied enzyme-catalyzed reactions, so his work borders onto biochemistry (Section 3.12). One of Prelog's main contributions concerns chiral molecules, i.e. molecules that have two forms differing from one another as the right hand does from the left. Stereochemically specific reactions have great practical importance, as many drugs, for example, are active only in one particular geometric form.

Organometallic compounds constitute a group of organic molecules containing one or more carbon-metal bond, and they are thus the organic counterpart to Werner's inorganic coordination compounds (see Section 3.6). In 1952 Ernst Otto Fischer (1918–) and Sir Geoffrey Wilkinson (1921–1996) independently described a completely new group of organometallic molecules, called sandwich compounds. In such compounds a metal ion is bound not to a single carbon atom but is 'sandwiched' between two aromatic organic molecules. Fischer and Wilkinson shared the Nobel Prize for Chemistry in 1973.

Work on the interaction of metal ions with organic molecules was also recognized by the prize in 1987, which was shared by Donald J. Cram (1919–) of UCLA, Jean-Marie Lehn (1939–) from Strasbourg (and Paris) and Charles J. Pedersen (1904–1989) of the Du Pont Company. These three investigators have synthesized molecules with a ring structure, in which the hole in their middle specifically recognizes and binds different metal ions. They can, for example, distinguish between closely related ions, such as those of sodium and potassium, and thus they mimic enzymes in their specificity. The first such compound was synthesized by Pedersen in 1967, and later Lehn and Cram developed increasingly sophisticated organic compounds with cavities and cages in which not only metal ions but other molecules are bound. This research has applications in the whole spectrum of the chemical field, from inorganic chemistry to biochemistry.

George A. Olah (1927–) from the University of Southern California was awarded the Nobel Prize for Chemistry in 1994 "for his contributions to carbocation chemistry." Already in the 1920s and 1930s chemists had suggested that positively charged ions of hydrocarbons are formed as short-lived intermediates in organic chemical reactions. Such carbocations were, however, thought to be so reactive and unstable that it would be impossible to prepare them in quantity. Olah's investigations, starting in the 1960s, contradicted this supposition, since he showed that stable carbocations can be prepared by the use of a new type of extremely acidic compounds ('superacids'), and carbocation chemistry now has a prominent position in all modern textbooks of organic chemistry.

The preparation of a new form of carbon compounds was also recognized by the Nobel Prize for Chemistry in 1996 to Robert F. Curl, Jr., (1933–) of Rice University, Sir Harold W. Kroto (1939–) of the University of Sussex and Richard E. Smalley (1943–) of Rice University. These investigators had in 1985 discovered compounds, called fullerenes, in which 60 or 70 carbon atoms are bound together in clusters in the form of a ball. The designation fullerenes is taken from the name of an American architect, R. Buckminster Fuller, who had designed a dome having the form of a football for 1967 Montreal World Exhibition.

3.8. *Preparative Organic Chemistry*

One of the chief goals of the organic chemist is to be able to synthesize increasingly complex compounds of carbon in combination with various other elements, such as hydrogen, oxygen, nitrogen, sulfur and phosphorus. The first Nobel Prize for Chemistry recognizing pioneering work in preparative organic chemistry was that to Victor Grignard (1871–1935) from Nancy and Paul Sabatier (1854–1941) from Toulouse in 1912. Grignard had discovered that organic halides can form compounds with magnesium. These compounds, now generally called Grignard reagents, are very reactive, and they are consequently widely used for synthetic purposes. Sabatier was given the prize for developing a method to hydrogenate organic compounds in the presence of metallic catalysts. With his method oils can be converted to saturated fats, and it is, for example, used for margarine production and other industrial processes.

The prize in 1950 was presented to Otto Diels (1876–1954) from Kiel and Kurt Alder (1902–1958) from Cologne "for their discovery and development of the diene synthesis," also called the Diels–Alder reaction. In this reaction, which was developed already in 1928, organic compounds containing two double bonds ('dienes') can effect the syntheses of many cyclic organic substances. During the decades following the original work several industrial applications of the Diels–Alder reaction have been found, for example, in the production of plastics, which may explain the lateness of the prize.

The German organic chemist Hans Fischer (1881–1945) from Munich had already done significant work on the structure of hemin, the organic pigment in hemoglobin, when he synthesized it from simpler organic molecules in 1928. He also contributed much to the elucidation of the structure of chlorophyll, and for these important achievements he was awarded the Nobel Prize for Chemistry in 1930 (cf. Section 3.5). He finished his determination of the structure of chlorophyll in 1935, and by the time of his death he had almost completed its synthesis as well.

Robert Burns Woodward (1917–1979) from Harvard is rightly considered the founder of the most advanced, modern art of organic synthesis. He designed methods for the total synthesis of a large number of complicated natural products, for example, cholesterol, chlorophyll and vitamin B_{12}. He received the Nobel Prize for Chemistry in 1965, and he would probably have received a second chemistry prize in 1981 for his part in the formulation of the Woodward–Hoffmann rules (see Section 3.4), had it not been for his early death. Work in synthetic organic chemistry was also recognized in 1979 with the prize to Herbert C. Brown (1912–) of Purdue University and Georg Wittig (1897–1987) from Heidelberg, who had developed the use of boron- and phosphorus-containing compounds, respectively, into important reagents in organic synthesis. Another master in chemical synthesis is Elias James Corey (1928–) from Harvard, who received the prize in 1990. He had made a brilliant analysis of the theory of organic synthesis, which permitted him to synthesize biologically active compounds of a complexity earlier considered impossible.

The Nobel Prize for Chemistry in 1984 was given to Robert Bruce Merrifield (1921–) of Rockefeller University "for his development of methodology for chemical synthesis on a solid matrix." Specifically, Merrifield applied this ingenious idea to the synthesis of large peptides and small proteins, for example, ribonuclease (cf. Section 3.12), but the principle has later also been applied to nucleic acid chemistry. In earlier methods each intermediate in the synthesis had to be isolated, which resulted in a drastic drop in yield in syntheses involving a large number of consecutive steps. In Merrifield's method these isolation steps are replaced by a simple washing procedure, which removes by-products as well as remaining starting materials, and in this way substantial losses are avoided.

3.9. *Chemistry of Natural Products*

The synthesis of complex organic molecules must be based on detailed knowledge of their structure. Early work on plant pigments was carried out by Richard Willstätter (1872–1942), a student of Adolf von Baeyer from

Munich (see Section 2). Willstätter showed a structural relatedness between chlorophyll and hemin, and he demonstrated that chlorophyll contains magnesium as an integral component. He also carried out pioneering investigations on other plant pigments, such as the carotenoids, and he was awarded the Nobel Prize for Chemistry in 1915 for these achievements. Willstätter's work laid the ground for the synthetic accomplishments of Hans Fischer (see Section 3.8). In addition, Willstätter contributed to the understanding of enzyme reactions.

The prizes for 1927 and 1928 were both presented to Heinrich Otto Wieland (1877–1957) from Munich and Adolf Windaus (1876–1959) from Göttingen, respectively, at the Nobel ceremony in 1928. These two chemists had done closely related work on the structure of steroids. The award to Wieland was primarily for his investigations of bile acids, whereas Windaus was recognized mainly for his work on cholesterol and his demonstration of the steroid nature of vitamin D. Wieland had already in 1912, before his prize-winning work, formulated a theory for biological oxidation, according to which removal of hydrogen (dehydrogenation) rather than reaction with oxygen is the dominating process.

Investigations on vitamins were recognized in 1937 and 1938 with the prizes to Sir Norman Haworth (1883–1950) from Birmingham and Paul Karrer (1889–1971) from Zürich and to Richard Kuhn (1900–1967) from Heidelberg. Haworth did outstanding work in carbohydrate chemistry, establishing the ring structure of glucose. He was the first chemist to synthesize vitamin C, and this is the basis for the present large-scale production of this nutrient. Haworth shared the prize with Karrer, who determined the structure of carotene and of vitamin A. Kuhn also worked on carotenoids, and he published the structure of vitamin B_2 at the same time as Karrer. He also isolated vitamin B_6. In 1939 the Nobel Prize for Chemistry was shared between Adolf Butenandt (1903–1995) from Berlin and Leopold Ružička (1887–1976) of ETH, Zurich. Butenandt was recognized "for his work on sex hormones," having isolated estrone, progesterone and androsterone. Ružička synthesized androsterone and also testosterone.

The awards for outstanding work in natural-product chemistry continued after World War II. In 1947 Sir Robert Robinson (1886–1975) from Oxford received the prize for his studies on plant substances, particularly alkaloids, such as morphine. Robinson also synthesized steroid hormones, and he elucidated the structure of penicillin. Many hormones are of a polypeptide nature, and in 1955 Vincent du Vigneaud (1907–1997) of Cornell University was given the prize for his synthesis of two such hormones, vasopressin and oxytocin. Finally, in this area, Alexander R. Todd (Lord Todd since 1962) (1907–1997) was recognized in 1957 "for his work on nucleotides and nucleotide co-enzymes." Todd had synthesized ATP (adenosine triphosphate) and ADP (adenosine diphosphate), the main energy carriers in living cells, and he determined the structure of vitamin B_{12} (cf. Section 3.5) and of FAD (flavin–adenine dinucleotide).

3.10. *Analytical Chemistry and Separation Science*

Inorganic chemists, organic chemists and biochemists develop analytical methods as part of their regular research. It is consequently natural that not many Nobel Prizes have been awarded for contributions specifically in analytical chemistry. One such prize was, however, that to Fritz Pregl (1869–1930) from Graz in 1923 for his development of organic microanalysis. The medical biochemist from Uppsala, Olof Hammarsten (1841–1932), who gave the presentation speech as Chairman of the Nobel Committee for Chemistry, stressed that Pregl's work constituted an improvement rather than a discovery, in accord with Nobel's will. Pregl modified existing methods for quantitative elemental analysis of organic substances to handle very small quantities, which saved time, labor and expense. Another prize in analytical chemistry was given to Jaroslav Heyrovsky (1890–1967) from Prague in 1959 for his development of polarographic methods of analysis. In these a dropping mercury electrode is employed to determine current-voltage curves for electrolytes. A given ion reacts at a specific voltage, and the current is a measure of the concentration of this ion.

The analysis of macromolecular constituents in living organisms requires specialized methods of separation. One such method is ultracentrifugation, developed by The Svedberg (1884–1971) from Uppsala a few years before he was awarded the Nobel Prize for Chemistry in 1926 "for his work on disperse systems" (see Section 3.11). Svedberg's student, Arne Tiselius (1902–1971), studied the migration of protein molecules in an electric field, and with this method, named electrophoresis, he demonstrated the complex nature of blood proteins. Tiselius also refined adsorption analysis, a method first used by the Russian botanist, Michail Tswett (1872–1919), for the separation of plant pigments and named chromatography by him. In 1948 Tiselius was given the prize for these achievements. A few years later (1952) Archer J. P. Martin (1910–) from London and Richard L. M. Synge (1914–1994) from Bucksburn (Scotland) shared the prize "for their invention of partition chromatography," and this method was a major tool in many biochemical investigations later awarded with Nobel Prizes (see Section 3.12).

3.11. *Polymers and Colloids*

Polymeric substances in solution, including life constituents, such as proteins and polysaccharides, are in a colloidal state, i.e. they exist as suspensions of particles one-millionth to one-thousandth of a centimeter in size. In the case of the biological polymers the individual molecules are so large that they form a colloidal suspension, but many other substances can be obtained in a colloidal state. A much-studied example is aggregates of gold atoms, and the Nobel Prize for Chemistry for 1925 was given to Richard Zsigmondy (1865–1929) from Göttingen for demonstrating the heterogeneous nature of such gold sols. He did this with the aid of an instrument, the ultramicroscope, which he had developed in collaboration with scientists at the Zeiss factory in Jena. With this instrument the particles and their motion can be observed by the light they scatter at a right angle to the direction of the illuminating light beam. Early work in colloid chemistry had also been carried out by Wolfgang Ostwald (1883–1943), son of the 1909 Laureate Wilhelm Ostwald, but this was not of a caliber earning him a Nobel Prize.

The Svedberg who received the Nobel Prize for Chemistry in 1926, also investigated gold sols. He used Zsigmond's ultramicroscope to study the Brownian movement of colloidal particles, so named after the Scottish botanist Robert Brown (1773–1858), and confirmed a theory developed by Albert Einstein (1859–1955) in 1905 and, independently, by M. Smoluchowski. His greatest achievement was, however, the construction of the ultracentrifuge, with which he studied not only the particle size distribution in gold sols but also determined the molecular weight of proteins, for example, hemoglobin. In the same year as Svedberg got the prize the Nobel Prize for Physics was awarded to Jean Baptiste Perrin (1870–1942) of Sorbonne for developing equilibrium sedimentation in colloidal solutions, a method which Svedberg later perfected in his ultracentrifuge. Svedberg's investigations with the ultracentrifuge and Tiselius's electrophoresis studies (see Section 3.10) were instrumental in establishing that protein molecules have a unique size and structure, and this was a prerequisite for Sanger's determination of their amino-acid sequence and the crystallographic work of Kendrew and Perutz (see Section 3.5).

In the 1920s Hermann Staudinger (1881–1965) from Freiburg developed the concept of macromolecules. He synthesized many polymers, and he showed that they are long chain molecules. The large plastic industry is largely based on Staudinger's work. In 1953 he received the Nobel Prize for Chemistry "for his discoveries in the field of macromolecular chemistry." The prize in 1963 was shared by Karl Ziegler (1898–1973) of the Max-Planck-Institute in Mülheim and Giulio Natta (1903–1979) from Milan for their discoveries in polymer chemistry and technology. Ziegler demonstrated that certain organometallic compounds (see Section 3.7) can be used to effect polymerization reactions, and Natta showed that Ziegler catalysts can produce polymers with a highly regular three-dimensional structure. Another Nobel Prize for contributions in polymer chemistry was given to Paul J. Flory (1910–1985) of Stanford in 1974. Flory carried out fundamental theoretical as well as experimental investigations of the physical chemistry of macromolecules, but his work also led to such important polymers as nylon and synthetic rubber. In 1977 a paper entitled "Syntheses of electrically conducting organic polymers: Halogen derivates of polyacetylene" was published in the *Journal of the American Chemical Society, Chemical Communications.* The authors of this paper, Alan J. Heeger (1936–) of the University of California at Santa Barbara, Alan G. MacDiarmid (1927–) of the University of Pennsylvania and Hideki Shirakawa (1936–) of the University of Tsukuba, Japan were awarded the Nobel Prize for Chemistry in 2000 for this discovery. The conducting polymers have already given rise to a number of applications such as photodiodes and light-emitting diodes and have future potential to generate microelectronics based upon plastic materials.

3.12. *Biochemistry*

The second Nobel Prize for discoveries in biochemistry came in 1929, when Sir Arthur Harden (1865–1940) from London and Hans von Euler-Chelpin (1873–1964) from Stockholm shared the prize for investigations of sugar fermentation, which formed a direct continuation of Buchner's work awarded in 1907. With his young co-worker, William John Young (1878–1942), Harden had shown in 1906 that fermentation requires a dialysable substance, called co-zymase, which is not destroyed by heat. Harden and Young also demonstrated that the process stops before all sugar (glucose) has been used up, but it starts again on addition of inorganic phosphate, and they suggested that hexose phosphates are formed in the early steps of fermentation. von Euler had done important work on the structure of co-zymase, shown to be nicotinamide adenine dinucleotide (NAD, earlier called DPN). As the number of Laureates can be three, it may seem appropriate for Young to have been included in the award, but Euler's discovery was published together with Karl Myrbäck (1900–1985), and the number of Laureates is limited to three.

The next biochemical Nobel Prize was given in 1946 for work in the protein field. James B. Sumner (1887–1955) of Cornell University received half the prize "for his discovery that enzymes can be crystallized" and John H. Northrop (1891–1987) together with Wendell M. Stanley (1904–1971), both of the Rockefeller Institute, shared the other half "for their preparation of enzymes and virus proteins in a pure form." Sumner had in 1926 crystalized an enzyme, urease, from jack beans and suggested that the crystals were the pure protein. His claim was, however, greeted with great scepticism, and the crystals were suggested to be inorganic salts with the enzyme adsorbed or occluded. Just a few years after Sumner's discovery Northrop, however, managed to crystalize three digestive enzymes, pepsin, trypsin and chymotrypsin, and by painstaking experiments shown them to be pure proteins. Stanley started his attempt to purify virus proteins in the 1930s, but not until 1945 did he get virus crystals, and this then made it possible to show that viruses are complexes of protein and nucleic acid. The pioneering studies of these three investigators form the basis for the enormous number of new crystal structures of biological macromolecules, which have been published in the second half of the 20th century (cf. Section 3.5).

Several Nobel Prizes for Chemistry have been awarded for work in photosynthesis and respiration, the two main processes in the energy metabolism of living organisms (cf. Section 3.5). In 1961 Melvin Calvin (1911–1997) of Berkeley received the prize for elucidating the carbon dioxide assimilation in plants. With the aid of carbon-14 (cf. Section 3.6) Calvin had shown that carbon dioxide is fixed in a cyclic process involving several enzymes. Peter Mitchell (1920–1992) of the Glynn Research Laboratories in England was awarded in 1978 for his formulation of the chemiosmotic theory. According to this theory, electron transfer (cf. Sections 3.3 and 3.4) in the membrane-bound enzyme complexes in both respiration and photosynthesis, is coupled to proton translocation across the membranes, and the electrochemical gradient thus created is used to drive the synthesis of ATP (adenosine triphosphate), the energy storage molecule in all living cells. Paul D. Boyer (1918–) of UCLA and John Walker (1941–) of the MRC Laboratory in Cambridge shared one-half of the 1997 Prize for their elucidation of the mechanism of ATP synthesis; the other half of the Prize went to Jens Skou (1918–) in Aarhus for the first discovery of an ion-transporting enzyme. Walker had determined the crystal structure of ATP synthase, and this structure confirmed a mechanism earlier proposed by Boyer, mainly on the basis of isotopic studies.

Luis F. Leloir (1906–1987) from Buenos Aires was awarded in 1970 "for the discovery of sugar nucleotides and their role in the biosynthesis of carbohydrates." In particular, Leloir had elucidated the biosynthesis of glycogen, the chief sugar reserve in animals and many microorganisms. Two years later the prize went with one half to Christian B. Anfinsen (1916–1995) of NIH and the other half shared by Stanford Moore (1913–1982) and William H. Stein (1911–1980), both from Rockefeller University, for fundamental work in protein chemistry. Anfinsen had shown, with the enzyme ribonuclease, that the information for a protein assuming a specific three-dimensional structure is inherent in its amino-acid sequence, and this discovery was the starting point for studies of the mechanism of protein folding, one of the major areas of present-day biochemical research. Moore and Stein had determined the amino-acid sequence of ribonuclease, but they received the prize for discovering anomalous properties of functional groups in the enzyme's active site, which is a result of the protein fold.

Naturally a number of Nobel Prizes for Chemistry have been given for work in the nucleic acid field. In 1980 Paul Berg (1926–) of Stanford received one half of the prize for studies of recombinant DNA, i.e. a molecule containing parts of DNA from different species, and the other half was shared by Walter Gilbert (1932–) from Harvard and Frederick Sanger (see Section 3.5) for developing methods for the determination of the base sequences of nucleic acids. Berg's work provides the basis of genetic engineering, which has led to the large biotechnology industry. Base sequence determinations are essential steps in recombinant-DNA technology, which is the rationale for Gilbert and Sanger sharing the prize with Berg. Sidney Altman (1939–) of Yale and Thomas R. Cech (1947–) of the University of Colorado shared the prize in 1989 "for their discovery of the catalytic properties of RNA." The central dogma of molecular biology is: DNA → RNA → enzyme. The discovery that not only enzymes but also RNA possesses catalytic properties have led to new ideas about the origin of life. The 1993 prize was shared by Kary B. Mullis (1944–) from La Jolla and Michael Smith (1932–2000) from Vancouver, who both have given important contributions to DNA technology. Mullis developed the PCR ('polymerase chain reaction') technique, which makes it possible to replicate millions of times a specific DNA segment in a complicated genetic material. Smith's work forms the basis for site-directed mutagenesis, a technique by which it is possible to change a specific amino-acid in a protein and thereby illuminate its functional role.

3.13. *Applied Chemistry*

A few Nobel Prizes for Chemistry have recognized contributions outside the conventional basic chemical fields. The prize in 1931 went to Carl Bosch (1874–1940) and Friedrich Bergius (1884–1949), both from Heidelberg, "for the invention and development of chemical high pressure methods." Bosch had modified Haber's method for ammonia synthesis (see Section 3.6) to make it suitable for large-scale industrial use. Bergius used high-pressure methods to prepare oil by the hydrogenation of coal, and Bosch, like Bergius working at the large concern I. G. Farben, later improved the procedure by finding a good catalyst for the Bergius process.

Work in agricultural and nutritional chemistry led to the award of Artturi Ilmari Virtanen (1895–1973) from Helsinki in 1945. The citation particularly stressed his development of the AIV method, so named after the inventor's initials. Virtanen had first carried out biochemical studies of nitrogen fixation by plants with the aim of producing protein-rich crops. He then found that the fodder could be preserved with the aid of a mixture of sulfuric and nitric acid (AIV acid).

Finally, basic work in atmospheric and environmental chemistry was recognized in 1995 with the prize to Paul Crutzen (1933–), from the Netherlands, working at Stockholm University and later at the Max-Planck-Institute in Mainz, Mario Molina (1943–) of MIT and F. Sherwood Rowland (1927–) of UC, Irvine. These three investigators have studied in detail the chemical processes leading to the formation and decomposition of ozone in the atmosphere. In particular, they have shown that the atmospheric ozone layer is very sensitive to emission chemicals produced by human activity, and these discoveries have led to international legislation.

4. Concluding Remarks

The first hundred years of Nobel Prizes for Chemistry give a beautiful picture of the development of modern chemistry. The prizes cover the whole spectrum of the basic chemical sciences, from theoretical chemistry to biochemistry, and also a number of contributions to applied chemistry.

From a quantitative point of view, organic chemistry dominates with no less than 25 awards. This is not surprising, since the special valence properties of carbon result in an almost infinite variation in the structure of organic compounds. Also, a large number of the prizes in organic chemistry were given for investigations of the chemistry of natural products of increasing complexity and thus are on the border to biochemistry.

As many as 11 prizes have been awarded for biochemical discoveries. Even if the first biochemical prize was already given in 1907 (Buchner), only three awards in this area came in the first half of the century, illustrating the explosive growth of biochemistry in recent decades (8 prizes in 1970–1997). At the other end of the chemical spectrum, physical chemistry, including chemical thermodynamics and kinetics, dominates with 14 prizes, but there has also been 6 prizes in theoretical chemistry. Chemical structure is another large area with 8 prizes, including awards for methodological developments as well as for the determination of the structure of large biological molecules or molecular complexes. Industrial chemistry was first recognized in 1931 (Bergius, Bosch), but many more recent prizes for basic contributions lie close to industrial applications, for example, those in polymer chemistry.

Science is a truly international undertaking, but the western dominance of the Nobel scene is striking. No less than 49 scientists in the United States have received the Nobel Prize for Chemistry, but the majority have been given the prize after World War II. The first US prize was awarded in 1915 (for 1914, Richards), and only two more Americans got the prize before 1946 (Langmuir in 1932, Urey in 1934). German chemists form the second most awarded group with 26 Laureates, but 14 of these received the prize before 1945. Of the 25 British investigators recognized, on the other hand, no less than 19 got the prize in the second half of the century. France has 7 Laureates in chemistry, Sweden and Switzerland 5 each, and the Netherlands and Canada 3. One prize winner each is found in the following countries: Argentina, Austria, Belgium, Czechoslovakia, Denmark, Finland, Italy, Norway and Russia.

Extrapolating the trend of the 20th century Nobel Prizes for Chemistry, it is expected that in the 21st century theoretical and computational chemistry will flourish with the aid of the expansion of computer technology. The study of biological systems may become more dominant and move from individual macromolecules to large interactive systems, for example, in chemical signaling and in neural function, including the brain. And it is to be hoped that the next century will witness a wider national distribution of Laureates.

Bibliography

Westgren, A., in *Nobel — The Man and His Prizes*, ed. Odelberg, W. (Elsevier, New York, 1972), pp. 279–385.

Kormos Barkan, D., *Walther Nernst and the Transition in Modern Physical Science* (Cambridge University Press, 1999).

Rife, P., *Lise Meitner and the Dawn of the Nuclear Age* (Birkhäuser, 1999).

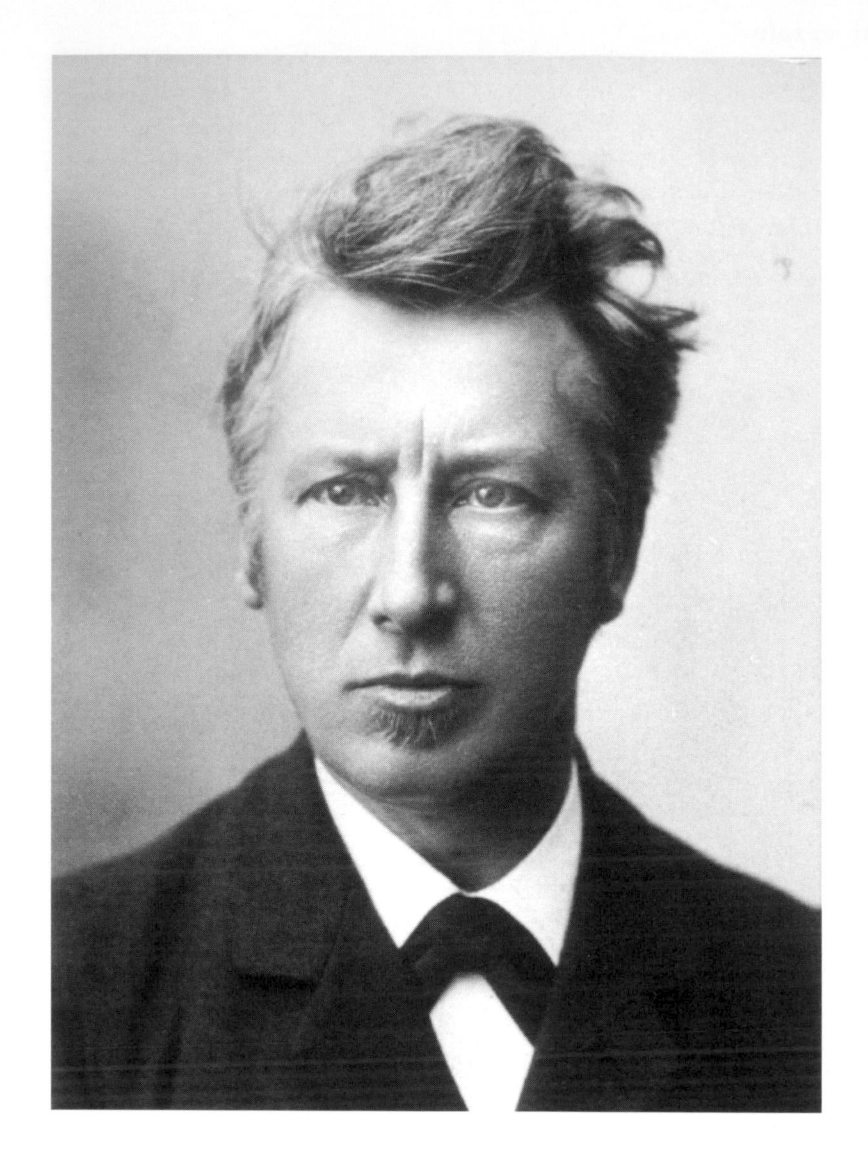

Chemistry 1901

Jacobus Henricus van't Hoff (1852–1911)

"in recognition of the extraordinary services he has rendered by the discovery of the laws of chemical dynamics and osmotic pressure in solutions"

Chemistry 1908

Ernest Rutherford (1871–1937)

"for his investigations into the disintegration of the elements, and the chemistry of radioactive substances"

Chemistry 1931

Carl Bosch (1874–1940)

"in recognition of contributions to the invention and development of chemical high pressure methods"

Chemistry 1944

Otto Hahn (1879–1968)

"for his discovery of the fission of heavy nuclei"

Chemistry 1954

Linus Carl Pauling (1901–1994)

"for his research into the nature of the chemical bond and its application to the elucidation of the structure of complex substances"

Chemistry 1958

Frederick Sanger (1918–)

"for his work on the structure of proteins, especially that of insulin"

Chemistry 1962

Max Ferdinand Perutz (1914–)

"for studies of the structures of globular proteins"

PHYSIOLOGY OR MEDICINE

"...one part to the person who shall have made the most important discovery within the domain of physiology or medicine..."

The Nobel Prize in Physiology or Medicine

Jan Lindsten and Nils Ringertz

Introduction

> "... The whole of my remaining realizable estate shall be dealt with in the following way: The capital shall be invested by my executors in safe securities and shall constitute a fund, the interest on which shall be annually distributed in the form of prizes to those who, during the *preceding year*, shall have conferred the *greatest benefit on mankind* ...; one part to the person who shall have made the most important *discovery* within the domain of *physiology or medicine*; ... The prizes for ... shall be awarded by ... that for physiology or medicine by *Carolinska Institute* in Stockholm; ..."

This is the exact wording of part of the translation into English of Alfred Nobel's will, which was signed in Paris on 27 November 1895. Together with the statutes of the Nobel Foundation, which were officially approved by the Swedish Government on 29 June 1900, the will constitutes the basis on which the Prize-Awarding Institutions execute their work.

Soon after the will became known to the public it was criticised for legal and other reasons. One complaint was that it was lacking in precision. The will was also contested by the children of Alfred Nobel's brothers, and once a settlement had been reached with the Nobel family, lengthy negotiations and compromises between the executors of the will and the Prize-Awarding Institutions were required before the statutes were approved by the Government. Even though the statutes introduced some practical rules they still left a number of points open for interpretation by the Prize-Awarding Institutions.

The aim of the present essay is to briefly review the Nobel Prize in Physiology or Medicine with regard to selection procedure and the discoveries which have been awarded Nobel Prizes. For information about the scientific work and the biographies of the Laureates the reader is referred to the series *Les Prix Nobel* which has been published every year since 1901 and most of which is now available online at www.nobelprize.org. (Comprehensive reviews covering the periods 1901–1950 and 1901–1960 were published by Liljestrand in 1950 and 1972.)

Selection of Nobel Laureates

Selection criteria

The Nobel Prize awarded by the Nobel Assembly at Karolinska Institutet is commonly referred to as the Nobel Prize in Medicine. The wording in Alfred Nobel's will, however, is *Physiology or Medicine.* It is important to make this distinction since, in the days of Alfred Nobel, physiology was used to describe what is today a number of biological fields. Interpreting the term 'physiology or medicine' in accordance with the intentions expressed in Alfred Nobel's will of 1895, therefore, leaves the Prize-Awarding Institution with considerable freedom to award Prizes in a broad biomedical field as well as in clinical medicine. Although the discussion of what is physiology and what is medicine is likely to continue indefinitely, it is clear that the prize awarder has on several occasions applied a broad definition. The prize in 1973 to Karl von Frisch, Konrad Lorenz and Nikolaas Tinbergen "for their discoveries concerning organisation and elicitation of individual and social behaviour patterns" could well be described as a prize in behavioral sciences. The prize in 1979 to Cormack and Hounsfield for "the development of computer-assisted tomography" fits well into the field of applied physics, while the prize in 1983 to Barbara McClintock for "the discovery of mobile genetic elements" could be considered a prize in plant genetics.

The key words in the will are *discovery* and *greatest benefit on mankind.* The essence of these words was thoroughly discussed during the preparatory work with the statutes of the Nobel Foundation. However, in the end no guidelines were provided. As a consequence it has been up to the Prize-Awarding Institution to interpret how these terms should be applied in the selection of Prize-winners. A further complication is that the corresponding wording is different for the two other Prizes within the natural sciences. In physics it is *discovery and invention* while in chemistry it is *discovery or improvement.*

Discoveries have been easier to define in the basic sciences than in clinical medicine. On the other hand "greatest benefit on mankind" has often been more obvious in clinical medicine than in basic research. Discovery has been defined as a sudden and significant increase in new knowledge rather than a steady increment of knowledge. As a consequence, awards have been given for scientific breakthroughs of high originality rather than for lifetime achievements. A discussion today of how the Prize-Awarding Institutions interpreted "greatest benefit on mankind" must take into account the knowledge available at the time of the award. Charles Nicolle received the prize in 1928 for "his work on typhus" which showed the role of the body louse in transmission of disease. Thanks to this discovery and simple delousing procedures it was possible to combat epidemics and save hundreds of thousands of lives during World War I. Paul Hermann Müller received the Nobel Prize in 1948 for his discovery of dichloro-diphenyl-trichloro-methylmethane (DDT) as an efficient insecticide. During World War II and immediately thereafter, DDT made it possible to master not only typhus epidemics, but also to combat other insect-transmitted diseases, such as malaria. The World Health Organization (WHO) estimates that during the period of its use approximately 25 million lives were saved. Today DDT is a banned chemical because of its environmental effects on fish and bird reproduction and its tendency to bioaccumulate in other species. However, at the time the prize was awarded, the benefit on mankind was quite obvious. Today, knowing about the disastrous ecological effects that the indiscriminate use of DDT as an insecticide in agriculture has had, the perspective is an entirely different one.

The will also states that the work recognized should have been carried out during the "preceding year." This rule has never been possible to follow in the strict sense. First, discoveries have to be made public in print which takes time. Second, they need to be verified by other researchers before becoming generally accepted. To solve this problem, the Prize-Awarders have been pragmatic and have interpreted "preceding year" to mean that the benefit of the discovery has become apparent to the Prize-Awarding Institution during the preceding year. The importance of this flexibility can be illustrated by the following example. Barbara McClintock made her first studies on mobile genetic elements in 1944, i.e. long before the structure of the DNA molecule was established by James D. Watson and Francis

H. C. Crick (1953). However, she was not awarded the Nobel Prize in Physiology or Medicine until 1983 (21 years after Watson and Crick). The main reason for this was that when the structure of DNA had been elucidated, and the genetic code had been deciphered by Nirenberg and Khorana (1968), it was obvious that changing even one single nucleotide in DNA could drastically disrupt reading of the genetic information. The scientific community simply found it very difficult to accept mobile genetic elements. Furthermore, it took a long time before it could be shown that mobile genetic elements is a general phenomenon and not a peculiar trait restricted to maize. Once other scientists had identified mobile genetic elements in bacteria and in insects, and it had become known that transposition of growth regulatory genes was involved in cancer, it was not long before Barbara McClintock was awarded the prize. The same line of reasoning is true for Peyton Rous who made his discovery of tumour viruses in chickens in 1916 but did not receive the prize until 50 years later when the existence of tumour viruses had been confirmed in other animal species.

One reason the will of Alfred Nobel was criticised was that it was too vague. Therefore, in formulating the statutes of the Nobel Foundation and of the Prize-Awarding Institutions, efforts were made to be more specific. A rule was added that *in no case may the prize be divided among more than three persons* (one-third to each Laureate, alternatively one-half to one of the Laureates and one-quarter to each of the other two). This criterion contributes to the exclusiveness of the prize, but it also imposes restrictions. It means that not all worthy candidates can be awarded the prize, and sometimes a whole scientific field has to be dropped because the number of candidates is too large and cannot be reduced to three in a logical and just way. On the other hand, it is not a scientific field that should be awarded but the scientist(s) who have made the most important discovery.

Sometimes an ethical problem arises when the death of a candidate in a combination of four opens the possibility to award the remaining three candidates. According to the statutes the Nobel Prize cannot be awarded posthumously.

The Nobel Prize in Physiology or Medicine has been shared more often during the second half of the twentieth century than during the first half. Between 1901 and 1950, 59 people received the Nobel Prize compared to 113 between 1951 and 2000. The increased tendency to split the prize most likely reflects the growth of the international scientific community after World War II and the fact that the number of people nominated each year has increased. Another plausible reason is that, increasingly, biomedical research is carried out by teams rather than by scientists working alone. But there are at least two other factors to consider. First, both World War I and II occurred during the first half of the twentieth century and Nobel Prizes were not awarded during some of the war years. Second, Jöns Johansson, the only member of the faculty who had worked with Alfred Nobel, was unhappy about the choice of some Prize-winners. For a number of years while on the Nobel Committee he effectively blocked the awarding of prizes in order to save money to build a Nobel research institute. The task of the institute would have been to assist the Nobel Committee by checking the accuracy of the discoveries claimed by the nominees.

Karolinska Institutet as a Nobel Prize-Awarding Institution

Karolinska Institutet awards the Nobel Prize in Physiology or Medicine, as stated in Alfred Nobel's will. The task was originally handled by the entire professorial staff which in 1901 comprised 19 members. The practical work was taken care of by a Nobel Committee comprised of three members, one of whom was the President of Karolinska Institutet and also the chairman of the committee. The first secretary of the Nobel Committee, Professor Göran Liljestrand, was not elected until 1918, and held the position for 42 years. During this period, the prestige of the Nobel Prize in Physiology or Medicine grew. Nevertheless, this long period of control by one person was one of several factors which later prompted changes in the organization.

In 1964 Karolinska Institutet had grown considerably and its research and teaching activities had become much more diversified. Selecting Nobel Laureates in physiology or medicine was the responsibility of a medical faculty which in 1970 counted 61 full professors. In addition, a university reform was planned which would include teachers other than professors (mainly associate professors and senior lecturers) in the faculty. After this reform the medical faculty would have had more than 200 members, thus making it difficult or impossible to conduct the complicated scientific discussions which represent a key element in the selection of Nobel Laureates. Furthermore, a new law was prepared in the 1970s implying that all documents at state institutions and organizations, of which Karolinska Institutet was one, would become open to the public. This would have made it impossible to keep secret the deliberations of the Nobel Committee. Since neither of these changes was considered to be beneficial for the

Nobel work, a new organization — the Nobel Assembly — was instituted in 1977.

The Nobel Assembly has very strong connections with Karolinska Institutet but, legally and financially, it is independent of the institute and the state. Its entire budget comes from the Nobel Foundation which handles all financial matters for the Prize-Awarding Institutions. The Nobel Assembly has 50 members, all of whom are active full professors at the institute. Members resign at the age of 65 years, i.e. when they retire from their positions at Karolinska Institutet. New members are elected by the assembly. A new chairman is elected every year according to seniority.

The Nobel Committee, which is the executive committee of the Nobel Assembly, is made up of five members and an executive secretary. In order to achieve a suitable balance between the need for both continuity and renewal, each committee member can only be elected for 3 + 3 consecutive years. The executive secretary can be elected for a maximum of four three-year periods. The mandates of the members begin and terminate in such a way that only part of the committee is renewed any single year. One of the members is elected chairman for a maximum of three years. After the committee of five has examined the nominations for a given year, an additional 10 ad hoc committee members are elected for a nine-month period. The inclusion of these members ensures that the committee has the necessary expertise to evaluate the work of the nominees who, judging from the nominations, are likely to be the top contenders for the prize in that particular year. The ad hoc members do not have to be members of the Nobel Assembly.

Several meetings between the assembly and the committee are held during the year in order to discuss the candidates nominated and the significance of their discoveries. Therefore, when it comes to the final decision, all assembly members know about the candidates nominated for the prize that year. During the first half of October, a decision is reached by the Nobel Assembly after voting by ballot (simple majority).

Nominations

The timetable for the prize has remained more or less the same since 1901. Thus, in September the year before the prize is to be awarded, confidential, personal *invitations to nominate candidates* for the prize are sent to 2500–3000 scientists who are members of medical faculties or academies outside Scandinavia. Scientists are invited according to a rotating system. Previous Nobel Laureates in Physiology or Medicine and professors at medical faculties in the Nordic countries have the right to nominate every year. Nominations are made on special forms sent only to those who are formally invited to nominate. The Nobel Committee receives many informal letters with invalid nominations. These are not included among the documents examined by the Nobel Committee.

The fact that it is the international scientific community that nominates the Prize candidates, and the principle that the invitations to nominate are rotated among different academies and individuals, contributes to confidence in the selection procedure and to the prestige of the Nobel Prizes. Furthermore, if someone is nominated several times over several years by different, independent scientists from different countries, this implies that the discovery made by the candidate has been widely recognized as being important. However, the total number of nominations in any particular year is not considered an important factor in judging the candidacy of a certain nominee. Nominations do not carry over from one year to another. To be considered for a particular year, the candidates must be nominated for that particular year irrespective of whether they have been previously nominated or not.

The deadline for the submission of nominations is 31 January (late incoming nominations are held over to the following year), after which the evaluation of the candidates begins.

All nominated candidates are evaluated by members of the Nobel Assembly and a written protocol is created for each new candidate. Preliminary reports and full reports on the work of individual nominees or, more commonly, groups of nominees, are prepared by members of the Nobel Assembly and by external reviewers. Top candidates are usually examined not once, but over a number of years, and by different reviewers. Each time the reviewer has access to the original nomination and all previous reports. Nowadays, all evaluations are made in English to make it possible for non-Swedish reviewers to access previous documentation. The evaluations have to be delivered before the end of August.

It is very unlikely that a candidate will receive the Nobel Prize in Physiology or Medicine the first year he or she is nominated, although this has happened on rare occasions, e.g. Carrel (1912), Hill and Meyerhof (1922) and Banting and Macleod (1923). It is worth mentioning that Banting and Macleod's first original publication on insulin appeared the year before they were awarded the prize, i.e. in 1922. One of the nominators of Banting and Macleod, Krogh, was almost as close. The discovery for which he was awarded the prize in 1920 was first published in Danish in

1918 and in English in 1919, and he was first nominated in 1919. There are some further examples of Laureates from the first 50 years who got the prize only one or two years after their first nomination, e.g. Domagk (1939), Fleming (1945) and Carl and Gerty Cori (1947). Their discoveries, however, had been made several years earlier.

The boundaries between Physics, Chemistry and Physiology or Medicine are not distinct. Thus, for instance, Röntgen was awarded the Nobel Prize in Physics but his discovery has been tremendously important not only in physics but also in chemistry and in medicine. Other Laureates in Physics have made discoveries which constitute a technological prerequisite for a considerable part of the research made within the biomedical field, especially after World War II. Schrödinger (Physics 1933) and Bohr (Physics 1922) played important roles in convincing biologists that life processes could be analysed in terms of atoms and molecules. By doing so they helped to create the field of molecular biology. Delbrück, a physicist who switched to biology, founded phage genetics, and won the Nobel Prize in Physiology or Medicine in 1969.

Laureates in Chemistry have made contributions which might equally well have been awarded a prize in Physiology or Medicine. Butenandt received the 1939 prize in Chemistry for his work on sex hormones, de Hevesy was awarded the 1943 Chemistry prize for introducing isotopes as tracers in the study of chemical processes, a technique of great value in biomedical research, while Sanger was awarded two Chemistry prizes for discoveries of great importance in modern biotechnology. In 1958 he received the Chemistry prize for elucidating the structure of insulin and then in 1980 he received a second prize for his method of sequencing nucleic acids. Other Nobel Prizes have been awarded for discoveries in the border zone between medicine and chemistry. Dorothy Hodgkin determined the structure of important biochemical substances by X-ray techniques and received the 1964 Chemistry prize, while Mitchell (Chemistry 1978) studied biological energy transfer and formulated chemiosmotic theory.

In the field of genetics some discoveries have been awarded prizes in Physiology or Medicine while others have received the Chemistry prize. The former group includes the 1933 prize to Morgan, the 1962 prize to Crick, Watson and Wilkins, and the 1993 prize to Roberts and Sharp. Molecular genetics has been recognized by prizes in Chemistry e.g. those to Berg, Gilbert, and Sanger (1980), Altman and Cech (1989), and Mullis and Smith (1993).

In order to coordinate the work of the Nobel Committees for Physiology or Medicine and Chemistry, a joint meeting of the committees is held in the spring. This also ensures that a candidate does not receive two Nobel Prizes for the same discovery.

The Laureates and Their Work

During the period 1901–2000 a total of 172 scientists were awarded the Nobel Prize in Physiology or Medicine. Their contributions range from basic to clinical research. The following is a compilation which gives an indication of the nature of the discoveries for which scientists were awarded Nobel Prizes. The prize motivations indicated are not verbatim the official ones but have been shortened to improve readability and overview (comments in parentheses are not part of the official prize sentence). The different prizes have been organized under different headings, but it should be pointed out that one discovery may fit under more than one heading.

Infectious agents and insecticides

Ross (1902)	role of insects as vectors in the infectious cycle (malaria)
Koch (1905)	identification of the tubercle bacillus and other work on tuberculosis
Laveran (1907)	role of protozoa in causing disease (malaria)
Nicolle (1928)	role of clothes lice in the transmission of typhus
Müller (1948)	DDT as an insecticide
Theiler (1951)	yellow fever virus
Enders, Weller and Robbins (1954)	*in vitro* culture of polio virus
Blumberg & Gajdusek (1976)	new mechanisms for the origin and dissemination of infectious disease
Prusiner (1997)	prions, a new principle of infection

Immunology

Behring (1901)	serum therapy and its application against diphteria
Ehrlich (1908)	immunity
Mechnikov (1908)	phagocytosis
Richet (1913)	anaphylaxis
Bordet (1919)	antigens and antibodies in immune reactions

Landsteiner (1930)	blood groups and blood typing
Burnet & Medawar (1960)	acquired immunological tolerance
Edelman & Porter (1972)	structure of antibodies
Benacerraf, Dausset & Snell (1980)	regulation of immune reactions
Jerne, Köhler & Milstein (1984)	control of the immune system and monoclonal antibodies
Tonegawa (1987)	genetics of antibody formation
Doherty & Zinkernagel (1996)	cell mediated immunity

Chemotherapy/drug development

Domagk (1939)	prontosil (sulphonamides)
Fleming, Chain & Florey (1945)	penicillin
Waksman (1952)	streptomycin, the first antibiotic against tuberculosis
Bovet (1957)	substances that mimic the effects of adrenalin. Substances that paralyse skeletal muscle.
Black, Elion & Hitchings (1988)	important principles for drug treatment

Phototherapy and fever treatment

Finsen (1903)	light therapy of lupus
Wagner–Jauregg (1927)	fever treatment of general paralysis

Cancer

Fibiger (1926)	spiroptera carcinoma illustrating cancer caused by chronic irritation
Rous (1966)	tumour-inducing virus in chickens
Huggins (1966)	hormonal treatment of prostate cancer
Baltimore, Dulbecco & Temin (1975)	interaction between tumour virus and host cell
Bishop & Varmus (1989)	retroviral oncogenes

Classical genetics

Morgan (1933)	role of chromosomes in heredity
Muller (1946)	production of mutations by X-ray irradiation
McClintock (1983)	mobile genetic elements

Cell biology

Claude, de Duve & Palade (1974)	structural and functional organisation of the cell
Cohen & Levi–Montalcini (1986)	growth factors
Blobel (1999)	the discovery that proteins have intrinsic signals that govern their transport and localisation in the cell

Developmental biology

Spemann (1935)	organiser effect in embryonic development
Lewis, Nüsslein–Volhard & Wieschaus (1995)	genetic control of early embryonic development

Molecular biology/genetics

Kossel (1910)	work on protein including the nucleic substances
Beadle & Tatum (1958)	regulation of definite chemical events (one gene–one protein)
Lederberg (1958)	genetic recombination and the organization of the genetic material in bacteria
Ochoa & Kornberg (1959)	mechanisms in the biological synthesis of ribonucleic and deoxyribonucleic acid
Crick, Watson & Wilkins (1962)	molecular structure of nucleic acids and its significance for information transfer in living material
Jacob, Lwoff & Monod (1965)	genetic control of enzyme and virus synthesis
Holley, Khorana & Nirenberg (1968)	genetic code and its function in protein synthesis
Delbrück, Hershey & Luria (1969)	the replication mechanism and the genetic structure of viruses
Arber, Nathans & Smith (1978)	restriction enzymes and their application to problems of molecular genetics
Roberts & Sharp (1993)	split genes

Intermediary metabolism

Hill (1922)	heat production in muscle

Meyerhof (1922)	oxygen consumption and the metabolism of lactic acid in muscle
Warburg (1931)	nature and mode of action of the respiratory enzyme
von Szent–Györgyi (1937)	vitamin C and the catalysis of fumaric acid
Cori & Cori (1947)	catalytic conversion of glycogen
Krebs (1953)	citric acid cycle
Lipmann (1953)	coenzyme A and its importance for intermediary metabolism
Theorell (1955)	nature and mode of action of oxidation enzymes
Bloch & Lynen (1964)	cholesterol and fatty acid metabolism
Bergström, Samuelsson & Vane (1982)	prostaglandins and related biologically active substances
Brown & Goldstein (1985)	regulation of cholesterol metabolism
Fischer & Krebs (1992)	reversible protein phosphorylation as a biological regulatory mechanism
Gilman & Rodbell (1994)	G-proteins and their role in signal transduction in cells

Hormones

Kocher (1909)	physiology, pathology and surgery of the thyroid gland
Banting & Macleod (1923)	insulin
Houssay (1947)	hormones of the anterior pituitary lobe in the metabolism of sugar
Kendall, Reichstein & Hench (1950)	hormones of the adrenal cortex, their structure and biological effects
Sutherland, Jr. (1971)	mechanism of action of hormones
Furchgott, Ignarro & Murad (1998)	nitric oxide as a signalling molecule in the cardiovascular system

Vitamins

Eijkman (1929)	antineuritic vitamin
Hopkins (1929)	growth-stimulating vitamin
Whipple, Minot & Murphy (1934)	liver therapy in cases of anaemia
Dam (1943)	vitamin K
Doisy (1943)	chemical nature of vitamin K

Digestion, circulation and respiration

Pavlov (1904)	physiology of digestion (conditioned reflexes)
Krogh (1920)	capillary motor regulating mechanism
Einthoven (1924)	electrocardiography
Heymans (1938)	role of sinus and aortic mechanisms in the regulation of respiration
Cournand, Forssmann & Richards (1956)	heart catheterization and pathological changes in the circulatory system

Neurobiology

Golgi & Ramón y Cajal (1906)	structure of the nervous system
Sherrington & Adrian (1932)	functions of neurones
Dale & Loewi (1936)	chemical transmission of the nerve impulses
Erlanger & Gasser (1944)	highly differentiated functions of single nerve fibres
Hess (1949)	functional organization of the interbrain as a coordinator of the activities of the internal organs
Bovet (1957)	synthetic compounds acting on the vascular system and skeletal muscles (curare)
Eccles, Hodgkin & Huxley (1963)	mechanisms involved in excitation and inhibition in the peripheral and central portions of the nerve cell membrane
Katz, von Euler & Axelrod (1970)	neurotransmittors and the mechanism for their storage, release and inactivation (the concept of synaptic transmission)
Guillemin & Schally (1977)	peptide hormone production in the brain
Yalow (1977)	radioimmunoassays of peptide hormones
Sperry (1981)	functional specialization of the cerebral hemispheres
Neher & Sakmann (1991)	function of single ion channels
Carlsson, Greengard & Kandel (2000)	signal transduction in the nervous system

Surgery

Moniz (1949)	therapeutic value of leucotomy in certain psychoses

Murray & Thomas (1990) — organ and cell transplantation in the treatment of human disease

Sensory physiology

Gullstrand (1911) — dioptrics of the eye
Bárány (1914) — physiology and pathology of the vestibular apparatus
von Békésy (1961) — physical mechanism of stimulation within the cochlea
Granit, Hartline & Wald (1967) — concerning the primary physiological and chemical visual processes in the eye
Hubel & Wiesel (1981) — information processing in the visual system

Behavioral sciences

von Frisch, Lorenz & Tinbergen (1973) — organization and elicitation of individual and social behavior patterns

Diagnostic methods

Cormack & Hounsfield (1979) — computer tomography

Geographic Distribution

"It is my express wish that in awarding the prizes no consideration whatever shall be given to the nationality of the candidates, but that the most worthy shall receive the prize, whether he be Scandinavian or not" (quote from the English translation of Alfred Nobel's will).

As the contents of Alfred Nobel's will became known the above sentence was considered unpatriotic and met with considerable criticism. With a perspective of 100 years of prizes, it is of some interest, therefore, to

	1901–1925	1926–1950	1951–1975	1976–2000
USA	1	13	32	40
Germany	5	3	3	4
UK	2	7	10	4
France	2	1	3	1
Others	13	12	9	7
Total	23	36	57	56

examine the geographical distribution of the recipients (as indicated by the domicile of the Laureates at the time of their awards).

Practically all Laureates come from or have carried out their work in Europe or North America. After World War II the United States has come to dominate the list. In the 1930s, the rise of the Nazis in Germany and Austria caused many Jewish scientists to flee from Europe to the United States. Among the 32 US scientists who received the Nobel Prize in Physiology or Medicine between 1951 and 1975 nine were born in countries other than the US. Of a total of 172 prize-winners, four are Danish and seven are Swedish. Other small countries have done equally well or better. Switzerland, with a population half that of Sweden and Denmark together, for instance has eight Nobel Laureates. Sweden and Switzerland were not involved in the two World Wars and could continue research relatively undisturbed while research work in other countries became impossible or difficult. The conclusion, however, is that there is no evidence that the prize-awarder has favored citizens of its own country.

Criticisms

There are three main types of criticism of the awarding of Nobel Prizes: omissions, mistakes and the failure to recognize the contributions of women. Since there are many prize-worthy candidates but only three who can receive the prize each year, omissions are bound to happen. The failure to award a Nobel Prize to Oswald T. Avery for the discovery of DNA as the genetic material can be used as one example. Avery undoubtedly discovered that DNA is the carrier of the genetic material when he showed that DNA from strains of bacteria with high pathogenicity could transform strains with low to high pathogenicity. His first publication on this topic appeared as early as 1944. Avery was nominated several times between 1932 and 1942 for work on polysacharide antigens. From 1945 he was nominated every year for his discovery concerning DNA. At the time many scientists thought of DNA, with its four different building blocks, as having too simple a structure to be the genetic material. Instead they favored the idea that proteins, with their 20 different amino acids, were more likely to be the genetic material and did not trust the enzyme digestions used by Avery to remove protein from his preparation of DNA. By the time the scientific community, including the Nobel Committee, had accepted Avery's data, he had passed away.

As far as mistakes are concerned, three examples are often brought up: Banting and Macleod (1923), Fibiger (1926) and Moniz (1949). The prize to Banting and Macleod for the discovery of insulin (1923) was questioned right from the time it was announced. Macleod was the department head and founder of the laboratory where Banting and a young colleague, Charles Best, worked. The discovery was made by Banting and Best at a time when Macleod was away. However, Banting, but not Best, was awarded the prize. This decision has been analysed in a monograph by Bliss (1982). The author arrived at the conclusion that the decision was the best one possible on the basis of what was known at the time. Others have emphasized the difficulties involved in trying to award prizes for discoveries made the preceding year. Incidentally, Best was first nominated in 1950.

The decision to award the 1926 prize to Fibiger for the discovery of the Spiroptera carcinoma has been heavily criticized. Very little was known about cancer-causing mechanisms at the time, and it was not until 40 years later that the next prize was awarded in the area of cancer research. At that time knowledge had accumulated about the genetic code, mutations, tumour viruses and other biological mechanisms involved in cancer.

The prize to Moniz for lobotomy (leucotomy) in 1949 must be seen in relation to the methods available for treating psychotic patients during the early part of the twentieth century. Before surgery was introduced, patients were subjected to very cruel treatments involving straitjackets and cold baths; then came surgical interventions and electric shock treatments. Both techniques became obsolete when neuropharmacology introduced effective drugs to treat psychosis. Today, lobotomy seems unethical but the question is whether it was unethical compared to the alternative methods available at the time of the award.

Only 6 of the 172 Nobel Laureates in Physiology or Medicine are women (Gerty T. Cori (1947); Rosalyn Yalow (1977); Barbara McClintock (1983); Rita Levi-Montalcini (1986); Gertrude B. Elion (1988); and Christiane Nüsslein-Volhard (1995)). This is not surprising, seen in the light of overwhelming male dominance within the biomedical field during the twentieth century.

Concluding Remarks

The Nobel Prizes awarded for Physiology or Medicine during the past century undoubtedly highlight a number of important discoveries. They do not, however, cover the whole story. There are a number of omissions and also a few prizes which, with hindsight, appear to be mistakes. The prestige of the Nobel Prize, however, comes from the fact that the international scientific community agrees with most of the decisions that have been made.

Alfred Nobel's intention was that the Prizes would make it possible for promising scientists to continue their research without having to worry about their financial situation. It is doubtful whether the Prize fulfils this function today. First, the average age at which prize-winners receive the prize is relatively high and, second, most of the recipients are already well established scientists when they receive the prize. Furthermore, medical research has gone the same way as physics and often demands considerable budgets and large research groups. The Nobel Prize, therefore, is not likely to make as big a difference today as it did a hundred years ago.

For the Nobel Assembly, the task of selecting the Nobel Laureates is a very stimulating one and the lectures of the Nobel Laureates are memorable occasions. The announcement of the Nobel Prize-winners in October as well as the award ceremony and festivities in December each year attract international media attention. This presents the Prize-Awarding Institutions with excellent opportunities to explain the achievements of the Laureates, to actively promote greater public understanding of science, and to interest young scholars in biomedical research.

Bibliography

Bliss, M. (1982) *The Discovery of Insulin* (McClelland and Stewart, Toronto).

Jansson, B. *Controversial Psychosurgery Resulted in a Nobel Prize*, Nobel e-Museum (www.nobel.se/medicine/articles/moniz/index.html).

Liljestrand, G. (1950) *The Prize in Physiology or Medicine*, in: Schück, H., Sohlman, R., Österling, A., Liljestrand, G., Westgren, A., Siegbahn, M., Schou, A. and Ståhle, N. K., *Nobel — The Man and His Prizes* (Sohlmans Förlag, Stockholm).

Liljestrand, G. (1972) *The Prize in Physiology or Medicine*, in: Schück, H., Sohlman, R., Österling, A., Liljestrand, G., Westgren, A., Siegbahn, M., Schou, A. and Ståhle, N. K., *Nobel — The Man and His Prizes,* Third revised, updated and enlarged edition (Elsevier Science Inc., New York).

Physiology or Medicine 1901

Emil Adolf von Behring (1854–1917)

"for his work on serum therapy, especially its application against diphtheria, by which he has opened a new road in the domain of medical science and thereby placed in the hands of the physician a victorious weapon against illness and deaths"

Physiology or Medicine 1933

Thomas Hunt Morgan (1866–1945)

"for his discoveries concerning the role played by the chromosome in heredity"

Physiology or Medicine 1962

Francis Harry Compton Crick (1916–)

"for discoveries concerning the molecular structure of nucleic acids and its significance for information transfer in living material"

Physiology or Medicine 1962

James Dewey Watson (1928–)

"for discoveries concerning the molecular structure of nucleic acids and its significance for information transfer in living material"

Physiology or Medicine 1965

François Jacob (1920–)

"for discoveries concerning genetic control of enzyme and virus synthesis"

Physiology or Medicine 1983

Barbara McClintock (1902–1992)

"for her discovery of mobile genetic elements"

Physiology or Medicine 1990

Joseph E. Murray (1919–)

"for discoveries concerning Organ and Cell Transplantation in the Treatment of Human Disease"

LITERATURE

"...one part to the person who shall have produced in the field of literature the most outstanding work in an ideal direction..."

The Nobel Prize in Literature

*Kjell Espmark**

1. Nobel's Will and the Literature Prize

Among the five prizes provided for in Alfred Nobel's will (1895), one was intended for the person who, in the literary field, had produced "the most outstanding work in an ideal direction." The Laureate should be determined by "the Academy in Stockholm," which was specified by the statutes of the Nobel Foundation to mean the Swedish Academy. These statutes defined literature as "not only belles-lettres, but also other writings which, by virtue of their form and style, possess literary value." At the same time, the restriction to works presented "during the preceding year" was softened: "older works" could be considered "if their significance has not become apparent until recently." It was also stated that candidates must be nominated in writing by those entitled to do so before 1 February each year.

A special regulation gave the right of nomination to members of the Swedish Academy and other academies, institutions and societies similar to it in constitution and purpose, and to university teachers of aesthetics, literature and history. An emendation in 1949 specified the category of teachers: "professors of literature and philology at universities and university colleges." The right to nominate was at the same time extended to previous Prize-winners and to "presidents of those societies of authors that are representative of the literary production in their respective countries." The

*Poet, novelist, and literary historian, former Professor in Comparative Literature at Stockholm University 1978–1995. Member of the Swedish Academy in 1981, Chairman of its Nobel Committee from 1988.

statutes also provided for a Nobel Committee "to give their opinion in matter of the award of the prizes" and for a Nobel Institute with a library which was to contain a substantial collection of mainly modern literature.

2. Accepting the Task? Discussion in the Swedish Academy

Two members of the Swedish Academy spoke strongly against accepting Nobel's legacy, for fear that the obligation would detract from the Academy's proper concerns and turn it into "a cosmopolitan tribunal of literature." They could have added that the Academy, in doldrums at the time, was ill-equipped for the sensitive task. The permanent secretary, Carl David af Wirsén, replied that refusal would deprive "the great figures of continental literature" of an exceptional recognition, and conjured up the weighty reproach to be directed at the Academy if it failed to "acquire an influential position in world literature." Besides, the task would not be foreign to the purposes of the Academy: proper knowledge of the best in the literature of other countries was necessary for an Academy that had to judge the literature of its own country. This effective argument, which won a qualified majority for acceptance, showed not only openness to Nobel's far-reaching intentions, but also harbored Wirsén's and his sympathizers' ambition to seize the unexpected possibilities in the field of the politics of culture, and to enjoy, as he wrote in a letter, "the enormous power and prestige that the Nobel will bequeaths to the Eighteen [members of the Academy]."

3. Nobel's Guidelines and Their Interpretations: A Short History

As guidelines for the distribution of the Literature Prize the Swedish Academy had the general requirement for all the prizes — the candidate should have bestowed "the greatest benefit on mankind" — and the special condition for literature, "in an ideal direction." Both prescriptions are vague and the second, in particular, was to cause much discussion. What did Nobel actually mean by ideal? In fact, the history of the Literature Prize appears as a series of attempts to interpret an imprecisely worded will. The consecutive phases in that history reflect the changing sensibility of an Academy continuously renewing itself. The main source of knowledge of the principles and criteria applied is the annual reports which the Committee presented to the Academy (itself making part of that body). Also the correspondence between the members is often enlightening. There is an obstacle though: all Nobel information is to be secret for 50 years.

3.1. '*A Lofty and Sound Idealism*'

The first stage, from 1901 to 1912, has the stamp of the secretary Carl David af Wirsén, who read Nobel's 'ideal' as 'a lofty and sound idealism'. The set of criteria which resulted in Prizes to Bjørnstierne Bjørnson, Rudyard Kipling and Paul Heyse, but rejected Leo Tolstoy, Henrik Ibsen and Émile Zola, is characterized by its conservative idealism (a domestic variation of Hegelian philosophy), holding church, state and family sacred, and by its idealist aesthetics derived from Goethe's and Hegel's epoch (and codified by F. T. Fischer in the middle of the nineteenth century). Those standards had earlier been typical of Wirsén's and the Academy's struggle against the radical Scandinavian writers. Nobel's testament gave Wirsén — called "the Don Quixote of Swedish romantic idealism" — the opportunity to carry his provincial campaign into the fields of international literature. This application was actually far from Nobel's values: he certainly shared Wirsén's disgust for writers like Zola, but was radically anticleric, adopting Shelley's utopian idealism and religiously coloured spirit of revolt.

3.2. *A Policy of Neutrality* (*World War I*)

The next chapter in the history of the Literary Prize could be entitled 'A Literary Policy of Neutrality'. The objectives laid down by the new chairman of the Academy's Nobel Committee at the beginning of the First World War kept, on the whole, the belligerent powers outside, giving the small nations a chance. This policy partly explains the Scandinavian overrepresentation on the list. The Prizes to the Swede Verner von Heidenstam, the Danes Karl Gjellerup and Henrik Pontoppidan — one of the few cases of a shared Prize — and to the Norwegian Knut Hamsun still in 1920 are to be comprehended from this point of view.

3.3. '*The Great Style*' (*the 1920s*)

A third period, approximately coinciding with the 1920s, could be labeled 'The Great Style'. This key concept in the reports of the Committee reveals the connections with Wirsén's epoch and its traits of classicism. With such a standard the Academy was, of course, out of touch with what happened in contemporary literature. It could appreciate Thomas Mann's *Buddenbrooks* — a masterpiece "approaching the classical realism in Tolstoy" — but passed his *Magic Mountain* over in silence. By that time, however, the Academy had got rid of its narrow definition of 'ideal direction'. In 1921 this stipulation of the will was interpreted more generously as 'wide-hearted humanity', which paved the way for writers like Anatole France and George Bernard Shaw, both inconceivable as Laureates — and, sure enough, rejected — at an earlier stage.

3.4. '*Universal Interest*' (*the 1930s*)

In line with the requirement "the greatest benefit on mankind," the Academy of the 1930s tried a new approach, equating this 'mankind' with the immediate readership of the works in question. A report of its Committee stated 'universal interest' as a criterion and the Academy decided on writers within everybody's reach, from Sinclair Lewis to Pearl Buck, repudiating exclusive poets like Paul Valéry and Paul Claudel.

3.5. '*The Pioneers*' (*1946–*)

Given a pause for renewal by the Second World War and inspired by its new secretary, Anders Österling, the post-war Academy finished this excursion into popular taste, focussing instead on what was called 'the pioneers'. Like in the sciences, the Laureates were to be found among those who paved the way for new developments. In a way, this is another interpretation of the formula "the greatest benefit on mankind": the perfect candidate was the one who had provided world literature with new possibilities in outlook and language.

In Österling's epoch, the word 'ideal' was deliberately taken in a still wider sense: the new list started with Hermann Hesse who, in the 1930s, had been rejected for 'ethical anarchy' and lack of 'plastic visuality and firmness' in his characters, words which echo Wirsén's time. Later, the compatibility of Samuel Beckett's dark conception of the world with Nobel's 'ideal' was put to the test, one of the last occasions when this condition was central to the discussion. It is only at 'the depths' that "pessimistic thought and poetry can work their miracles," said Karl–Ragnar Gierow in his address, emphasising the deep sense of human worth and the life-giving force, nevertheless, in Beckett's pessimism. The borderline of this generosity can be seen in the handling of Ezra Pound. He appealed to the Academy because of his 'pioneering significance', but was disqualified by his wartime applauding, on the Italian radio network, of the mass extermination of the East European Jews. Member Dag Hammarskjöld, in a representative way, concluded that "such a 'subhuman' reaction" excluded "a prize that is after all intended to lay weight on the 'idealistic tendency' of the recipient's efforts." (This repudiation did not prevent Hammarskjöld from negotiating, on the Academy's commission, with the American authorities for Pound's release from the mental hospital where he had been interned to be saved from a death penalty for treason.)

This new policy, at the same time more exclusive and more generous in its interpretation of the will, was actually meant to start with Valéry but he died in the summer of 1945. Instead we find, in 1946–1950, the splendid series Hesse, André Gide, T. S. Eliot, and William Faulkner. In his address to the author of *The Waste Land*, Österling drew attention to "another pioneer work, which had a still more sensational effect on modern literature," James Joyce's *Ulysses*. With this reference to the greatest omission of the 1930s, he extended the 1948 acclaim of Eliot to cover also the dead master. The explicit concentration on innovators can, via the choices of Saint-John Perse in 1960 and Samuel Beckett in 1969, be traced up to recent years.

The criterion lost weight, however, as the heroic period of the international avant-garde turned into history and literary innovation became less ostentatious. Instead, the instruments pointed at the 'pioneers' of specific linguistic areas. The 1988 Prize was awarded to a writer who, from a Western point of view, rather administers the heritage from Flaubert and Thomas Mann. In the Arabic world, on the other hand, Naguib Mahfouz appears as the creator of its contemporary novel. The following Prize went to Camilo José Cela, who had, in an international perspective, modest claims to the title 'pioneer', but who was, in Spanish literature, the great innovator of post-war fiction. Still found among these innovators of certain linguistic areas is the 2000 Laureate, Gao Xingjian, whose œuvre "has opened new paths for the Chinese novel drama."

3.6. *Attention to Unknown Masters (1978–)*

Another policy, partly coinciding with the one just outlined, partly replacing it, is "the pragmatic consideration" worded by the new secretary, Lars Gyllensten, and, again, taking into account the 'benefit' of the Prize. A growing number within the Academy wanted to call attention to important but unnoticed writers and literatures, thus giving the world audience masterpieces they would otherwise miss, and, at the same time, giving an important writer due attention. We get glimpses of such arguments as far back as the choice of Rabindranath Tagore in 1913 but there was no programme until the early 1970s. The full emergence of this policy can be seen from 1978 and onwards, in the Prizes to Isaac Bashevis Singer, Odysseus Elytis, Elias Canetti, and Jaroslav Seifert. The criterion gives poetry a prominent place. In no other period were the poets so well provided for as in the years 1990–1996 when four of the seven prizes went to Octavio Paz, Derek Walcott, Seamus Heaney, and Wisława Szymborska, all of them unknown earlier to the world audience.

3.7. '*The Literature of the Whole World*' (*1986–*)

A new policy, long on its way, had a breakthrough in the 1980s. Again, it was an attempt to understand and carry out Nobel's intentions. His will had an international horizon, though it rejected any consideration for the nationality of the candidates: the most worthy should be chosen, "whether he be a Scandinavian or not." The problem of surveying the literature of the whole world was, however, overwhelming and for a long time the Academy was, with justice, to be criticized for making the award a European affair. Wirsén expressly confined himself, as we saw, to "the great figures of Continental literature." In the 1920s it was certainly laid down that the prize was "intended for the literature of the whole world" but instruments to implement the idea were not available. In the 1930s, there were, on the whole, not even reasonable nominations from the Asiatic countries and the Academy had, at that time, not yet developed a scouting system of its own.

The Prize at last to Yasunari Kawabata in 1968 illustrates the exceptional difficulties in judging literature in non-European languages — this was a matter of seven years, involving four international experts. In 1984, however, Gyllensten declared that attention to non-European writers was gradually increasing in the Academy; attempts were being made "to achieve a global distribution." This includes measures to strengthen the competence for the international task.

The picture of the Academy's Eurocentric policy was also significantly altered by the choices of Wole Soyinka from Nigeria in 1986 and Naguib Mahfouz from Egypt in 1988. Later practice shows the extention to Nadine Gordimer from South Africa, to Kenzaburo Oe from Japan, to Derek Walcott from St. Lucia in the West Indies, to Toni Morrison, the first Afro–American on the list, and to Gao Xingjian, the first laureate to write in Chinese. It is, however, important that nationality is not involved in the discussion. It has sometimes been suggested that the Academy should first decide upon a neglected language and then seek out the best candidate in it. Doing so would amount to politization of the Prize. Instead, efforts are being made to widen the horizon so that, in the course of the normal process of judgement, it is possible to weigh sometimes a prominent Nigerian dramatist and poet, sometimes an Egyptian novelist, against candidates from closer parts of the linguistic atlas — with all such evaluations continuing to be made on literary grounds. Critics have quite often neglected the Academy's striving for political integrity. Naturally, an international prize can have political effects but it must not, according to this jury, carry any political intention.

The criteria discussed sometimes alternate, sometimes coincide. The spotlight on the unknown master Canetti in 1981 is thus followed by the laurel to the universally hailed 'pioneer' of magic realism, Gabriel García Márquez, in 1982. Some Laureates answer both requirements, like Faulkner, who was not only "the great experimentalist among twentieth-century novelists" — the Academy was here fortunate enough to anticipate Faulkner's enormous importance to later fiction — but also, in 1950, a fairly unknown writer. On this occasion, the Prize, for once, could help a great innovator outside the limelight to reach his potential disciples as well as his due audience. The surprising Prize to Dario Fo in 1997 can also be said to have a double address: it was given to a genre which had earlier been left out in the cold but also to the brilliant innovator of that genre.

3.8. *The Prize Becoming a Literary Prize*

The more and more generous interpretation of the formula "in an ideal direction" continued in the 1980s and the 1990s. Academy Secretary Lars Gyllensten pointed out that nowadays the expression "is not taken too literally ... It is realized that on the whole the serious literature that is worthy of a prize furthers knowledge of man and his condition and endeavours to enrich and improve his life." Cela's candidature, again, put the principle to the test. His dark conception of the world posed the same problem as Beckett's, and provoked a similar solution. The Prize was given "for a rich and intense prose, which with restrained compassion forms a challenging vision of man's vulnerability." As Knut Ahnlund said in his address, Cela's work "in no way lacks sympathy or common human feeling, unless we demand that those sentiments should be expressed in the simplest possible way." In this 'unless' we glimpse the repudiation, implicit in recent practice, of the early narrow interpretation of the will. The Nobel Prize in Literature has gradually become a literary prize. One of the few reminiscences of the 'ideal direction' policy of the earlier age is the homage paid to those great artistic achievements that are characterized by uncompromising 'integrity' in the depiction of the human predicament (cf. below).

3.9. *International Neglect of the Change of Standards*

International criticism of the Literature Prize has usually treated the Academy's practice during the first century of the Prize as a whole, overlooking the differences in outlook and criteria between the various periods, even neglecting the continuous renewal which makes the Academy of, say, 1950 a jury much different from Wirsén's.

As to the early prizes, the censure of bad choices and blatant omissions is often justified. Tolstoy, Ibsen and Henry James should have been rewarded instead of, for instance, Sully Prudhomme, Eucken and Heyse. The Academy which got this exacting commission was simply not fit for the task. It was deliberately formed as 'a bulwark' against the new radical literature in Sweden and much too conservative in outlook and taste to be an international literary jury. It was not until the 1940s — with Anders Österling as secretary — that the Academy, considerably rejuvenated, had the competence to address the major writers of, in the first place, the Western World. On the whole, criticism of its postwar practice has also been much more appreciative. Objections in recent times have less often been levelled against literary quality, rather referred, mistakenly, to political intentions. Also blame for eurocentricity was common, in particular from Asiatic quarters, up to the choices of Soyinka and Mahfouz in the 1980s.

4. Special Articles

4.1. *Nomination*

In the first year, the number of nominations was 25. In the early time of the Prize the members of the Swedish Academy were reluctant to use their right to nominate candidates. Impartiality suggested that proposals should come from outside. As no one abroad nominated Tolstoy in 1901, the self-evident candidate of the time fell outside the discussion. The omission caused a strong reaction from Swedish writers and artists who sent an address to Tolstoy — who answered by declining any future prize. During the First World War the number of nominations decreased, to fall to twelve in 1919, compared with 28 in 1913. This wartime slackening of initiative from the outside world induced the Academy to make use of its right to propose. In 1916 the Committee members themselves put forward five names. In recent times, members of the Committee — but also other members of the Academy — regularly add their nominations to the outside names to make the list as comprehensive and representative as possible. The number of nominations has towards the end of the century been about — and even substantially surpassed — 200.

4.2. *The Nobel Committee*

The Nobel Committe is a working unit of 3–5, chosen within the Swedish Academy (with a rare additional member from outside). Its task is to examine the proposals made and study all relevant literary material to select the candidates to be considered by the Academy. Formerly the Committee presented only one name for the decision of the Academy, which usually confirmed the choice of its Committee. (There are exceptions though: the Academy preferred Tagore in 1913 and Henri Bergson in 1927.) From the 1970s onward, the members of the Committee have presented individual reports, which enables the Academy to weigh the different opinions and consequently gives it a greater influence.

The Committee's first task is to trim down 'the long list' nowadays of about 200 names to some 15, which are presented to the Academy in April. Towards the end of May, this 'half-long list' is condensed to a 'short list' of five names. The œuvres of these finalists make up the Academy's summer readings. At its first reunion in the middle of September, the discussion immediately starts, to end in a decision about a month later. Naturally, the whole production of five writers would be too heavy a workload for a couple of months but most names of the previous short list return the current year, which makes the task more reasonable. It should be added that in recent times a first-year candidate will not be taken to a prize the same year. In the background looms one of the main failures, Pearl Buck, the Laureate of 1938. A first-year candidate, she was launched by a Committee minority as late as 19 September, to win the contest a short time afterwards, without due consideration.

The chairman of the Committee has usually been identical with the Academy's permanent secretary, with some displacement at transitional stages. Thus, Carl David af Wirsén was chairman in 1900–1912, Per Hallström (secretary from 1931) in 1922–1946, Anders Österling (secretary from

1941) in 1947–1970, Karl-Ragnar Gierow (secretary from 1964) in 1970–1980, and Lars Gyllensten (secretary from 1977) in 1981–1987. An exceptional period is in 1913–1921 when the historian, Harald Hjärne, wrote the reports. In 1986, when Sture Allén became secretary, Gyllensten remained as chairman, to be succeeded by Kjell Espmark in 1988. Since 1986 the tasks have thus been divided between secretary and chairman.

4.3. *'Ideal' — A Textual Examination*

As was shown by Sture Allén, the adjective 'ideal' referring to an ideal was used by several of Nobel's contemporaries; one of them was Strindberg. However, the word is, he found, an amendment made by Nobel in his handwritten will. Nobel seems to have written 'idealirad', with 'idealiserad' (idealized) in mind, but checked himself in front of the reference to embellishment in this word for upliftment and wrote 'sk' over the final letters 'rad', thus ending in the disputed word 'idealisk'. Allén concluded that Nobel actually meant "in a direction towards an ideal", and specified the sphere of the ideal by the general criterion for all the Nobel Prizes: they are addressed to those who "shall have conferred the greatest benefit on mankind." "This means, for instance," Allén added, "that writings, however brilliant, that advocate, say, genocide, will not comply with the will."

4.4. *Shared Prize*

The Nobel Prize for Literature can be divided between two — but not three — candidates. However, the Swedish Academy has been restrictive on this point. Divisions are liable to be regarded as — and sometimes are — the result of compromise. That was the case with Frédéric Mistral and José Echegaray in 1904 and with Karl Gjellerup and Henrik Pontoppidan in 1916. A shared prize also runs the risk of being viewed as only half a laurel. Later divisions are exceptional, the only cases being the shared Prizes to Shmuel Yosef Agnon and Nelly Sachs in 1966 and to Eyvind Johnson and Harry Martinson in 1974. In the 1970s a policy was laid down, stating (1) that each of the two candidates must alone be worthy of the Prize and (2) that there must be some community between them justifying the procedure. The latter requirement no doubt offers a real obstacle for divisions.

4.5. *Competence for the International Task*

In the Swedish Academy, linguistic competence has, as a rule, been high. French, English, and German have posed no problems and several members have been excellent translators from Italian and Spanish. Also noted Orientalists have found a place in the Academy. One of them (Esaias Tegnér, Jr.) could have read Tagore in Bengali (but in fact contented himself with the author's own English translation of Gitanjali), another (H. S. Nyberg) could report on Arabic literature. In 1985 Göran Malmqvist, one of the West's foremost experts on modern Chinese literature, became a member. The present Academy includes competence also in Russian. Above all, however, the area of scrutiny has been extended by means of specialists in the various fields. Where translations into English, French, German or the Scandinavian languages are missing, special translations can also be procured. In several cases such exclusive versions — with no more than eighteen readers — have played an important role in the recent work of the Academy.

4.6. '*Political Integrity*'

The Literary Prize has often, in particular during the cold war, given rise to discussion of its political implications. The Swedish Academy, for its part, has on many occasions expressed a desire to stand apart from political antagonisms. The guiding principle, in Lars Gyllensten's words, has been 'political integrity'. This has quite often not been understood. Especially in the East it has been hard to grasp the Swedish Academy's autonomous position vis-à-vis state and government. In fact, the Academy does not receive any subsidy from the state, nor would it accept any interference in its work. The government, in its turn, is quite happy to stand outside the delicate Nobel matters.

Naturally, there is a political aspect of any international literary prize. It is, however, necessary to make a distinction between political effects and political intentions. The former are unavoidable — and often unpredictable. The latter are expressly banned by the Academy. The distinction, as well as the autonomy of the Academy, can be illustrated by the prehistory of the Prize to Solzhenitsyn. Considering the sad consequences for Pasternak of his Prize, the secretary Karl–Ragnar Gierow took the unusual step of writing to the Swedish ambassador to Moscow, Gunnar Jarring, to gain some idea of Solzhenitsyn's position, stressing that the question related, of course, only to what might "happen to him personally." On this point, Mr. Jarring could give a reassuring answer (which proved not to be prophetic). But he also had another message. He wanted to postpone the decision, specifying, in a letter to Österling, that a prize to Solzhenitsyn "would lead to difficulties for our relations with the Soviet Union." He received the reply: "Yes, that could well be so, but we all agreed that Solzhenitsyn is the most deserving candidate." This exchange illuminates a fundamental fact: the Academy has no regard for what may or may not be desirable in the eyes of the Swedish Foreign Office. Its unconventional inquiry was concerned solely with the likely effects of the decision for the candidate personally. However, the exchange also offers a good example of the way in which a likely political effect may be taken into account — not, of course, that the Academy intended the possible disturbance in Soviet relations, but that it was aware of the risk and chose to take it.

The history of the Literary Prize offers a case where this delicate balance was endangered, the prize to Winston Churchill. When the decision was taken in 1953, after many years of discussion, it was felt that a sufficient distance from the candidate's wartime exploits had been gained, making it possible for a Prize to him to be generally understood as a literary award. The reaction from many quarters showed that this was quite a vain hope.

Now, there can be no doubt that the Committee and the Academy attributed exceptional literary merits to Churchill the historian and the orator. They certainly concurred in the address to the Laureate, "a Caesar who also had the gift of wielding Cicero's stylus." The problem was how this Caesar, a mere eight years after the war, could be mentally separated from the Ciceronian prose. After all, Churchill was not only the winner of World War II but prime minister and leader of one of the key powers in the cold war world. It can be asked if any of the Academy's choices has put its political integrity at such risk. At any rate, one well-known conclusion was drawn: ever since, candidates with governmental positions, such as André Malraux and Léopold Senghor, have been consistently ruled out.

During the last decades there is one seeming case of a 'political' Prize, the award to Czesław Miłosz. "Has Miłosz been given the 1980 Prize because Poland is politically in fashion?", asked *Der Tagesspiegel* and many other newspapers joined in. The suspicions did not account for the time involved in each nominee's candidacy. As was disclosed by a member, Artur Lundkvist, Miłosz had been on the list for three or four years and had been shortlisted in May 1980 — in other words, long before the Danzig strike. The strike caused several members to hesitate, said Lundkvist, but he added that it would have been equally impossible to drop Miłosz because of the events in Poland.

His argument no doubt reflects the opinion within the Academy. This jury realizes not only the damage that a political choice would inflict on the Prize; the integrity of the award could be jeopardised also by a non-choice in a delicate situation. Still, Miłosz was a dissident, and so were Jaroslav Seifert and Joseph Brodsky, the Laureates of 1984 and 1987. These choices all caused great irritation in the East. There one failed to see that the Academy's overriding concern was literary. The pronouncements of the secretary repeatedly stressed the existential dimensions of these great contemporary poets, values corresponding to the humanistic traditions of the Literary Prize. From that point of view it is essential that Miłosz's political defection be thus formulated by Gyllensten (after a reminder of how during the cold war the political climate had altered in a Stalinist direction): "With his uncompromising demand for artistic integrity and human freedom, Miłosz could no longer support the regime." Uncompromising integrity and a call to rally round human values — these are qualities that the Swedish Academy, following the spirit of Nobel's will, has again and again sought in combination with great artistic achievement. And just as repeatedly, this mode of evaluation has collided with Marxist/Leninist aesthetics, which interprets such a focus as mere camouflage for political intentions.

The process of judgement, while 'primarily a literary matter', does not, of course, prevent subsidiary evaluations from gradually forming a pattern. Such a pattern is apparent in the sequence Singer–Miłosz–Canetti–Seifert. At first sight one could see here what a newspaper headline proclaimed about the choice of Seifert: "The Swedish Academy Greets Central Europe." It is, however, not a question of some politically defined region or some third way in the tug-of-war between East and West. It is rather a question of authors who with great personal integrity have given voice to an old culture that has either been swept aside by oppressors or whose continued existence was severely threatened. In the difficult area of Central Europe, a number of authors have emerged, speaking, out of their sorely tested experience, on behalf of the basic human values — this in keeping with the humanistic tradition of the Nobel Prize. Such a pattern, though, reveals only part of the truth. The Prize is in the end not given to an attitude toward life, to a set of cultural roots, or to the substance of a commitment; the Prize has been rewarded so as to honor the unique artistic power by which this human experience has been shaped into literature.

4.7. *International Criticism of the Literature Prize*

The history of the Literature Prize is also the history of its reception in the press and in other media. Apart from overlooking the changes in outlooks and criteria within the Swedish Academy, international criticism has tended to neglect the crowd of likely names around the Prize for a specific year. Thus, Graham Greene was a celebrated candidate towards 1970 and the Academy was criticized for passing him over. But the 1969 Prize went to Samuel Beckett and the 1970 Prize to Aleksandr Solzhenitsyn, both most worthy candidates. Quite rightly, an international inquiry by *Books Abroad* in 1951, directed to 350 specialists, came to the conclusion that the first fifty years of the Prize contained 150 'necessary' candidates. The Academy cannot have the ambition to crown all worthy writers. What it cannot afford is giving Nobel's laurel to a minor talent. Its practice during the last full half-century has also largely escaped criticism on that point. Even the inquiry of 1951 found that two-thirds of the prizes during the first half-century were fully justified — "a fairly decent testimonial," as Österling commented. The second half-century as liable to get a still better mark.

As was mentioned above, criticism of omissions and bad choices was often justified as to the early period of the Prize. The Academy headed by Wirsén made only one choice to get general acclaim by posterity — Rudyard Kipling, and then for qualities other than those that have shown themselves to be lasting. The score of the 1910s and the 1920s was better: Gerhart Hauptmann, Tagore, France, Yeats, Shaw, and Mann have been found worthy in several appraisals. The results of the period 1930–1939 are poorer. Two choices have widely been regarded as splendid: Luigi Pirandello in 1934 and Eugene O'Neill in 1936. But the period offers several laureates justly judged as mediocre — and they conceal as many cases of neglect: Virginia Woolf ought to have been rewarded instead of Pearl Buck, and so on. The Academy of the inter-war years quite simply lacked the necessary tools to evaluate one of the most dynamic periods in Western literature. The post-war Academy has in a quite different manner fulfilled the expectations of serious criticism. The Österling Academy's investment in the pioneers has received due recognition in many favorable assessments. Names like Gide, Eliot, Faulkner, Hemingway, and Beckett have won general acclaim. Some names less known to an international audience, like Jiménez, Laxness, Quasimodo, and Andrić, have attracted criticism as insignificant, but been classified by experts as discoveries.

Sometimes the complaints about omissions have been anachronistic. Among those missing, critics have found Proust, Kafka, Rilke, Musil, Cavafy,

Mandelstam, García Lorca, and Pessoa. This list, if it had any chronological justification, would undeniably suggest serious failure. But the main works of Kafka, Cavafy, and Pessoa were not published until after their deaths and the true dimensions of Mandelstam's poetry were revealed above all in the unpublished poems that his wife saved from extinction and gave to the world long after he had perished in his Siberian exile. In the other cases there was much too brief a period of time between the publication of the author's most deserving work and his death for a prize to have been possible. Thus, Proust achieved notoriety in 1919 by the Goncourt Prize for the second part of *À la recherche du temps perdu* but less than three years later he was dead. The same short time of reaction was offered by Rilke's *Duineser Elegien* and García Lorca's plays. Musil's significance did not appear outside a narrow circle of connoisseurs until more than a decade after his death in 1942. He belonged, as was pointed out by a critic (Theodor Ziolkowski), to the category of authors who "on closer examination ... exclude themselves."

5. Epilogue: At the Turn of the Century

The last literary Nobel Prize of the twentieth century was awarded to Günter Grass, "whose frolicsome black fables portray the forgotten face of history." The choice won general acclaim but the moment was called in question. Why not three decades ago when Grass was at the summit of his craft? And why just now?

The first question takes us back to the situation around 1970 when Böll and Grass were both hot names. When the laurel was given to Böll in 1972 the citation recalled his contribution "to a renewal of German literature." The word had, however, a special meaning here. As was clarified in Gierow's speech to the Laureate 'the renewal' was 'not an experiment with form' but 'a rebirth out of annihilation', 'a resurrection' of a ravaged culture 'to the joy and benefit of us all': "Such was the kind of work Alfred Nobel wished his prize to reward." This meant that the foremost representative of a moral renaissance from the ruins of the Third Reich was preferred, with a direct appeal to Nobel's intentions, to the country's foremost representative of what was an artistic renewal. The choice took Grass out of focus for many years, and allowed for a discussion of a downward trend in his craft. It remained for the rejuvenated Academy of the nineties to take up the issue again. Several of its new members might have chosen Grass instead of Böll in 1972. As to the alleged decline of

Grass's art, the presentation at the announcement certainly called special attention to *The Tin Drum* and the Danzig trilogy it makes part of, but refused to share the politically biased German view of *Ein weites Feld*. "We just read the book and it is goddam good," as the permanent secretary Horace Engdahl declared.

Also the second question — why just now? — can be answered. The citation recalls the fabulous historian, with a view to the forgotten face of history. Without neglecting works like *The Flounder*, beginning at the dawn of history, the jury naturally focused upon the great recreator of the century just about to end. Grass is, in the secretary's words, "one of the really important writers investigating and explaining the twentieth century to us"; giving him the last prize of the century was 'an easy decision'. In other words, the choice long due found its perfect moment at the very end of the period that Grass had summed up in his incomparable way.

Grass's stronger position in recent years is, of course, also due to the growing understanding of his role as a source of energy in literature. In 1972 he was still a solitary master. In recent years he has been hailed as a precursor by writers such as Salman Rushdie, Nadine Gordimer, Gabriel García Márquez, Antonio Lobo Antunes, and Kenzaburo Oe. Grass has found his place among the 'pioneers'.

This choice at the end of the century has, however, also another purport. The Prizes to Hesse, Gide, Eliot, and Faulkner introduced a half-century of new competence for the difficult mission. The 1999 Prize is an indication of how far the jury has managed to make the Prize for Literature a literary award. The reference to moral values at the expense of experimental art in 1972 would be hard to imagine in the present Academy. We also notice the explicit disregard of the political implications that made Grass's last novel an apple of discord in his country. The Literary Prize has made an instructive journey since 1901. At the beginning of the new century it has become the Literary Prize that its name announces.

Bibliography

Espmark, K., *The Nobel Prize in Literature. A Study of the Criteria behind the Choices*, G.K. Hall & Co, Boston, 1991.

Literature 1901

Sully Prudhomme
(pen-name of René François Armand Prudhomme) (1839–1907)

"in special recognition of his poetic composition, which gives evidence of lofty idealism, artistic perfection and a rare combination of the qualitites of both heart and intellect"

Literature 1913

Rabindranath Tagore (1861–1941)

"because of his profoundly sensitive, fresh and beautiful verse, by which, with consummate skill, he has made his poetic thought, expressed in his own English words, a part of the literature of the West"

Literature 1925

George Bernard Shaw (1856–1950)

"for his work which is marked by both idealism and humanity, its stimulating satire often being infused with a singular poetic beauty"

Literature 1929

Thomas Mann (1875–1955)

"principally for his great novel, 'Buddenbrooks', which has won steadily increased recognition as one of the classic works of contemporary literature"

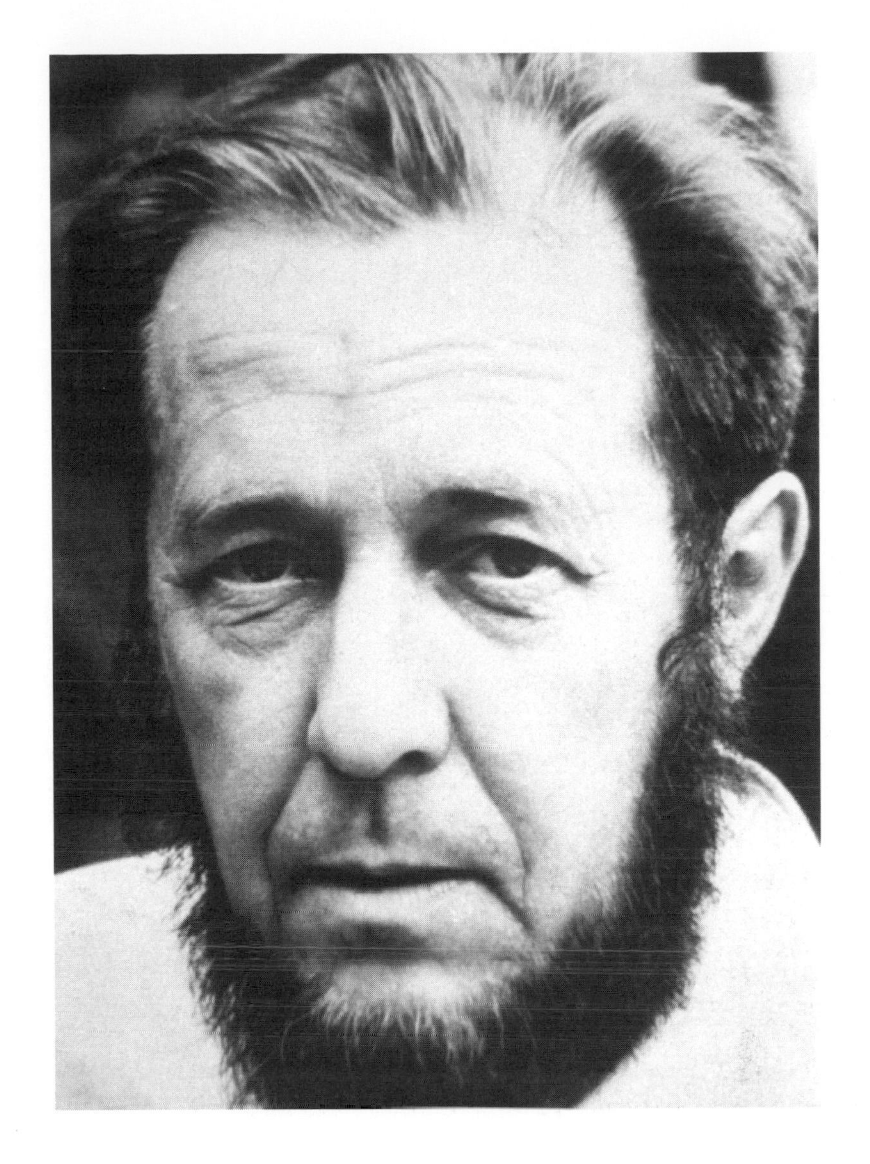

Literature 1970

Aleksandr Isaevich Solzhenitsyn (1918–)

"for the ethical force with which he has pursued the indispensable traditions of Russian literature"

Literature 1993

Toni Morrison (1931–)

"who in novels characterized by visionary force and poetic import, gives life to an essential aspect of American reality"

Literature 1996

Wisława Szymborska (1923–)

"for poetry that with ironic precision allows the historical and biological context to come to light in fragments of human reality"

PEACE

"...one part to the person who shall have done the most or the best work for fraternity between nations, for the abolition or reduction of standing armies and for the holding and promotion of peace congresses..."

The Nobel Peace Prize

*Geir Lundestad**

1. Introduction

This article is intended to serve as a basic survey of the history of the Nobel Peace Prize during its first 100 years. Since all the 107 Laureates selected from 1901 to 2000 are to be mentioned, the emphasis will be on facts and names. At the same time, however, I shall try to deal with two central questions about the Nobel Peace Prize. First, why does the Peace Prize have the prestige it actually has? Second, what explains the nature of the historical record the Norwegian Nobel Committee has established over these 100 years?

There are more than 300 peace prizes in the world. None is in any way as well known and as highly respected as the Nobel Peace Prize. *The Oxford Dictionary of Twentieth Century World History*, to cite just one example, states that the Nobel Peace Prize is "The world's most prestigious prize awarded for the 'preservation of peace'." Personally, I think there are many reasons for this prestige: the long history of the Peace Prize; the fact that it belongs to a family of prizes, i.e. the Nobel family, where all the family members benefit from the relationship; the growing political independence of the Norwegian Nobel Committee; the monetary value of the prize, particularly in the early and in the most recent years of its history. In this context, however, I am going to concentrate on the historical record of the Nobel Peace Prize. In my opinion, the prize would never have enjoyed

*Director of the Norwegian Nobel Institute in Oslo and Secretary of the Norwegian Nobel Committee since 1990.

the kind of position it has today had it not been for the decent, even highly respectable, record the Norwegian Nobel Committee has established in its selections over these 100 years. One important element of this record has been the committee's broad definition of peace, enough to take in virtually any relevant field of peace work.

On the second point, the selections of the Norwegian Nobel Committee reflected the insights primarily of the committee members and secondarily of its secretaries and advisors.

But, on a deeper level, they also generally reflected Norwegian definitions of the broader, Western values of an idealist, the often slightly left-of-center kind, but rarely so far left that the choices were not acceptable to Western liberal–internationalist opinion in general. The Norwegian government did not determine the choices of the Norwegian Nobel Committee, but these choices reflected the same mixture of idealism and realism that characterized Norwegian, and Scandinavian, foreign policy in general. As we shall see, some of the most controversial choices occurred when the Norwegian Nobel Committee suddenly awarded prizes to rather hard-line realist politicians.

2. Nobel's Will and the Peace Prize

When Alfred Nobel died on December 10, 1896, it was discovered that he had left a will, dated November 27, 1895, according to which most of his vast wealth was to be used for five prizes, including one for peace. The prize for peace was to be awarded to the person who "shall have done the most or the best work for fraternity between nations, for the abolition or reduction of standing armies and for the holding of peace congresses." The prize was to be awarded "by a committee of five persons to be elected by the Norwegian Storting."

Nobel left no explanation as to why the prize for peace was to be awarded by a Norwegian committee while the other four prizes were to be handled by Swedish committees. On this point, therefore, we are dealing only with educated inferences. These are some of the most likely ones: Nobel, who lived most of his life abroad and who wrote his will at the Swedish–Norwegian Club in Paris, may have been influenced by the fact that, until 1905, Norway was in union with Sweden. Since the scientific prizes were to be awarded by the most competent, i.e. Swedish, committees at least the remaining prize for peace ought to be awarded by a Norwegian committee. Nobel may have been aware of the strong interest of the

Norwegian Storting (Parliament) in the peaceful solution of international disputes in the 1890s. He might have in fact, considered Norway a more peace-oriented and more democratic country than Sweden. Finally, Nobel may have been influenced by his admiration for Norwegian fiction, particularly by the author Bjørnstjerne Bjørnson, who was a well-known peace activist in the 1890s. Or it may have been a combination of all these factors.

While there was a great deal of controversy surrounding Nobel's will in Sweden and that of the role of the designated prize awarding institutions, certainly including the fact that the rebellious Norwegians were to award the Peace Prize, the Norwegian Storting quickly accepted its role as awarder of the Nobel Peace Prize. On April 26, 1897, a month after it had received formal notification from the executors of the will, the Storting voted to accept the responsibility, more than a year before the designated Swedish bodies took similar action. It was to take three years of various legal actions before the first Nobel Prizes could actually be awarded.

3. 1901–1913: The Peace Prize to the Organized Peace Movement

Although there was nothing in the statutes that prevented the Storting from naming international members, the members of the Nobel Committee of the Storting (as the committee was called until 1977) have all been Norwegians from the very beginning. They were selected by the Storting to reflect the strengths of the various parties, but the members elected their own chairman. From December 1901 and until his death in 1922, Jørgen Løvland was the chairman of the Nobel Committee. He was one of the leaders of the Venstre (Left) party and served briefly as Foreign Minister (1905–1907), and then as Prime Minister (1907–1908). A majority of the five committee members in this period consistently represented that party.

Initially, Venstre represented a broad democratic-nationalist coalition, emphasizing universal suffrage, first for men, later for women, and independence from Sweden. The party strongly wanted to isolate Norway from Great Power politics; not only did it want Norway's full independence, but also some form of guaranteed permanent neutrality, based on the Swiss model. Yet at the same time, the party had a definite interest in international peace work in the form of mediation, arbitration and the peaceful solution of disputes. Small countries, certainly including Norway, were to show the world the way from Great Power politics to a world based on law and norms.

Norwegian parliamentarians, particularly from Venstre, took a strong interest in the Inter-Parliamentary Union formed in 1889. After Switzerland, Norway was the first country to pledge an annual contribution, first for its general operations (1895), and then for its office in Bern (1897). Norway was to have hosted the Union's conference in 1893, but because of the tense situation vis-à-vis Sweden the conference in Oslo was held only in 1899. These same liberal politicians were also highly sympathetic to the peace groups and societies that sprang up in many countries in the last decades of the 1800s, groups which starting in 1889 were internationally organized in the more or less annual Universal Peace Congress. The Permanent International Peace Bureau, founded in 1891 in Bern, became the international headquarters of this popular movement. (The movement long struggled with difficult finances, despite small fixed annual grants from Switzerland, Denmark, Sweden and Norway.) A third element in the peace work of this period was the more official movement, culminating in the Hague Conferences of 1899 and 1907, called by Tsar Nicholas II of Russia, to the enormous surprise of the governments of most other major powers. (The tsar was never seriously considered for the Peace Prize.)

Those few members of the Nobel Committee who did not represent Venstre tended to be jurists who took a special interest in building peace through international law, a desire shared by Venstre. Thus, former conservative Prime Minister and law professor Francis Hagerup was a committee member from 1907 to 1920. He was also chairman of the Norwegian delegation to the second Hague Conference in 1907.

With this composition of the Nobel Committee in mind, the list of the Nobel Laureates for the years 1901 to 1914 comes as no big surprise. Of the 19 prizes awarded during this period, only two went to persons who did not represent the Inter-Parliamentary Union, popular peace groups or the international legal tradition. The first two elements may also be said to have reflected the point in Nobel's will about the prize being awarded for "the holding and promotion of peace congresses."

The first prize in 1901 was awarded to Frédéric Passy (and Jean Henry Dunant). Passy was an obvious choice for the first prize since he had been one of the main founders of the Inter-Parliamentary Union and also the main organizer of the first Universal Peace Congress. He was himself the leader of the French peace movement. In his own person, he thus brought together the two branches of the international organized peace movement, the parliamentary one and the broader peace societies.

In 1902, the Peace Prize was awarded to Élie Ducommun, veteran peace advocate and the first honorary secretary of the International Peace Bureau, and to Charles Albert Gobat, first Secretary General of the Inter-Parliamentary Union and who later became Secretary General of the International Peace Bureau. (In 1906–1908 Gobat coordinated both groups, further underlining the close relationship between them.) In 1903 the prize went to William Randal Cremer, the 'first father' of the Inter-Parliamentary Union. In 1889, Bertha von Suttner had published her anti-war novel *Lay Down Your Arms.* After that, she was drawn into the international peace movement. She undoubtedly exercised considerable influence on Alfred Nobel, whom she had known since 1876, when he later decided to include the Peace Prize as one of the five prizes mentioned in his will. In 1905, she was awarded the Peace Prize, the first woman to receive such a distinction. Her supporters strongly felt that the prize had come too late, since she had had such an influence on Nobel. In 1907, the prize was awarded to Ernesto Teodoro Moneta, a key leader of the Italian peace movement. In 1908, the prize was divided between Fredrik Bajer, the foremost peace advocate in Scandinavia, combining work in the Inter-Parliamentary Union with being the first president of the International Peace Bureau, and Klas Pontus Arnoldson, founder of the Swedish Peace and Arbitration League. In 1910, the Permanent International Peace Bureau itself received the prize. In 1911, Alfred Hermann Fried, founder of the German Peace Society, leading peace publisher/educator and a close collaborator, shared it with Tobias Michael Carel Asser. In 1913, Henri La Fontaine was the first socialist to receive the Nobel Peace Prize. He was head of the International Peace Bureau from 1907 until his death in 1943. He was also active in the Inter-Parliamentary Union.

International legal work for peace represented the third road to the Nobel Peace Prize. In 1904, the Institute of International Law, the first organization or institution to receive the Peace Prize, was honored for its efforts as an unofficial body to formulate the general principles of the science of international law. In 1907, Louis Renault, leading French international jurist and a member of the Permanent Court of Arbitration at The Hague, shared the Peace Prize with Ernesto Teodoro Moneta. In 1909, the prize was shared between Paul Henri Benjamin Balluet, Baron d'Estournelles de Constant de Rebecque, who combined diplomatic work for Franco–German and Franco–British understanding with a distinguished career in international arbitration, and Auguste Marie François Beernaert, former Belgian Prime Minister, representative to the two Hague conferences, and a leading figure in the Inter-Parliamentary Union. Like d'Estournelles and Renault, Beernaert was also a member of the Permanent Court of Arbitration. Thus, few if any of the Laureates summed up the different stands of the early peace movement in the way Beernaert did. The Laureate of 1911, Tobias Michael Carel Asser, was also a member of the Court of Arbitration as well as the initiator of the Conferences on International Private Law. When America's Elihu Root received the Nobel Peace Prize in 1912, he had served both as U.S. Secretary of War and Secretary of State. But he was awarded the prize primarily for his strong interest in international arbitration and for his plan for a world court, which was finally established in 1920.

Jean Henry Dunant (1901) and Theodore Roosevelt (1906) are the two Laureates who clearly fall outside any of the categories mentioned so far. Dunant, who founded the International Red Cross in 1863, had been more or less forgotten until a campaign secured him several international prizes, including the first Nobel Peace Prize. The Norwegian Nobel Committee thus established a broad definition of peace, arguing that even humanitarian work embodied 'the fraternity between nations' that Nobel had referred to in his will. Roosevelt was the twenty-sixth president of the United States and the first in a long series of statesmen to be awarded the Nobel Peace Prize. He received the prize for his successful mediation to end the Russo–Japanese war and for his interest in arbitration, having provided the Hague arbitration court with its very first case. Internationally, however, he was best known for a rather bellicose posture, which certainly included the use of force. It is known that both the secretary and the relevant adviser of the Nobel Committee at that time were highly critical of an award to Roosevelt. It is thus tempting to speculate that the American president was honored at least in part because Norway, as a new state on the international arena, "needed a large, friendly neighbor — even if he is far away," as one Norwegian newspaper put it. Even if, or perhaps rather because, the prize to Roosevelt was controversial, it did in some ways constitute a breakthrough in international media interest in the Nobel Peace Prize.

4. 1914–1918: The First World War and the Red Cross

The First World War signified the collapse of the peaceful world which so many of the peace activists honored by the Nobel Peace Prize had worked so hard to establish. During the war, the number of nominations for the prize diminished somewhat, although a substantial number was still put forward. During the difficult war years, the Nobel Committee in neutral Norway decided to award no prize, except the one in 1917 to the

International Committee of the Red Cross (ICRC). The ICRC had been established in 1863 as a Swiss committee; the preceding year, the Convention for the Amelioration of the Conditions of the Wounded in Armies in the Field (the Geneva Convention) had been signed. During the First World War, the ICRC undertook the tremendous task of trying to protect the rights of the many prisoners of war on all sides, including their right to establish contacts with their families.

5. 1919–1939: The League of Nations and the Work for Peace

In the 1920s, Venstre's domination of the Nobel Committee continued even after the death of Jørgen Løvland and despite the choice of Conservative law professor Fredrik Stang (1922–1941) as the new chairman of the committee and the inclusion of Labor party historian Halvdan Koht in 1919. Old-timers Hans Jakob Horst and Bernhard Hanssen served on the committee from 1901 to 1931 and from 1913 to 1939, respectively. They were joined by Johan Ludwig Mowinckel (1925–1936) who meanwhile served as both Norway's Prime Minister and Foreign Minister during three separate periods. In the 1930s, the membership of the committee became more mixed, but the Venstre members now maintained the balance between more conservative and social democratic members. Still, during the period from 1919 to 1939, the growing political tension within the committee and the presence of certain stubborn individuals resulted in as many as nine 'irregular' years, when either no prize was awarded or it was awarded one year late, compared to only one such year during the period from 1901 to 1913.

After the First World War, Norway became a member of the League of Nations. This break with the past was smaller than it might seem. In the Storting, 20 members, largely Social Democrats, voted against membership. Even most of the 100 who voted in favor came to insist on the right to withdraw from the sanctions regime of the League in case of war. Norway basically still perceived itself as a neutral state. The old ideals of mediation, arbitration, and the establishment of international legal norms definitely survived, only slightly tempered by the experiences of the war and the membership in the League. Yet at the same time, some states and some statesmen were definitely regarded as better than others. Most Norwegian foreign policy leaders felt closest to Great Britain and the United States, despite significant fishery disputes with the former and the geographical distance and isolationism of the latter.

At least eight of the 21 Laureates in the period from 1919 to 1939 had a clear connection with the League of Nations. For the Nobel Committee the League came to represent the enhancement of the Inter-Parliamentary Union tradition from before 1914. In 1919, the Peace Prize was awarded to the President of the United States, Thomas Woodrow Wilson for his crucial role in establishing the League. Wilson had been nominated by many, including Venstre Prime Minister Gunnar Knudsen. In a certain sense the prize to Wilson was obvious; what still made it controversial, also among committee members, was that the League was part of the Versailles Treaty, which was regarded as diverging from the president's own ideal of 'peace without victory'. The prize in 1920 to Léon Victor Auguste Bourgeois, a prominent French politician and peace activist, showed the continuity between the pre-1914 peace movement and the League. Bourgeois had participated in both the Hague Conferences of 1899 and 1907; in 1918–1919 he pushed for what became the League to such an extent that he was frequently called its 'spiritual father'.

Swedish Social Democratic leader Karl Hjalmar Branting had also done long service for peace, but was particularly honored in 1921 with the Peace Prize for his work in the League of Nations. His fellow Laureate, Norway's Christian Lous Lange, the first secretary of the Norwegian Nobel Committee, had been the secretary-general of the Inter-Parliamentary Union since 1909 and had done important work in keeping the Union alive even during the war. After the war he was active in the League until his death in 1938. In 1922, the Norwegian Nobel Committee honored another Norwegian, Fridtjof Nansen, for his humanitarian work in Russia, which was done outside the League, but even more importantly for his work on behalf of the League to repatriate a great number of prisoners of war. From 1921, he was the League's High Commissioner for Refugees. The refugee problem proved rather intractable. The Nansen International Office for Refugees was authorized by the League in 1930 and was closed only in 1938. For its work, it received that year's Nobel Peace Prize.

In 1934, British Labour leader Arthur Henderson received the Peace Prize for his work for the League, particularly its efforts in disarmament. No single individual was more closely identified with the League from its beginning to its end than Viscount Cecil of Chelwood, who was honored with the prize in 1937. Only Koht's threat of resignation from the committee prevented the Peace Prize from being awarded directly to the League of Nations. In 1924, the committee even discussed awarding the prize to the Inter-Parliamentary Union.

In the years 1919–1939, the Nobel Committee also continued to honor the less official workers for peace. Since the peace societies of the pre-1914 period had lost most of their importance, this category of Laureates was now considerably more mixed than it had been in the earlier period. The clearest connection to the past was found in the shared prize for 1927 to Ludwig Quidde and Ferdinand Buisson. Buisson had joined his first peace society as early as 1867 and he had also been active in the Inter-Parliamentary Union, while the younger Quidde had joined the German Peace Society in 1892. In 1927, they were honored for their contributions to Franco–German popular reconciliation. In 1930, Lars Olof Nathan Söderblom was the first church leader to be awarded the Nobel Peace Prize for his efforts to involve the churches not only in work for ecumenical unity, but also for world peace. In 1931, Jane Addams was honored for her social reform work, but even more for establishing in 1919, and then leading, the Women's International League for Peace and Freedom (WILPF). Sharing the prize with Jane Addams was Nicholas Murray Butler, president of the Carnegie Endowment for International Peace, promoter of the Briand–Kellogg pact and leader of the more establishment–oriented part of the American peace movement. Sir Norman Angell, who received the Peace Prize for 1933, had written his famous book *The Great Illusion* as early as 1910. In the book he argued that war did not pay, not that it was impossible as it was frequently understood to have stated. In the inter-war years, he was a strong supporter of the League of Nations as well as an influential publicist/educator for peace in general.

The most clear-cut representative in this period of the legal tradition to limit or even end war was the former American Secretary of State, Frank Billings Kellogg. He was awarded the 1929 Peace Prize for the Kellogg–Briand pact, whose signatories agreed to settle all conflicts by peaceful means and renounced war as an instrument of national policy.

While Theodore Roosevelt and, to a lesser extent, Elihu Root, were the only prominent international politicians to receive the Nobel Peace Prize in the years before 1914, at least five prominent politicians in addition to Kellogg were to be so honored between 1919 and 1939. In 1926 alone, the Nobel Committee actually awarded the reserved prize for 1925 to Vice President Charles Gates Dawes of the United States and Foreign Secretary Sir Austen Chamberlain of Great Britain and the 1926 Prize to Foreign Minister Aristide Briand of France and Foreign Minister Gustav Stresemann of Germany. Dawes was responsible for the Dawes Plan for German reparations which was seen as having provided the economic underpinning of the Locarno Pact of 1925, under which Germany accepted its western borders as final. The four prizes reflected recognition of the changed international political climate, particularly between Germany and France, which Locarno helped bring about. It was probably also an effort by the committee to strengthen Norway's relations with the four international powers that mattered most for its interests. In 1936, the prize was awarded to Argentine Foreign Minister Carlos Saavedra Lamas for his mediation of an end to the Chaco War between Paraguay and Bolivia. Lamas also played a significant role in the League of Nations.

The most controversial award of the inter-war period was undoubtedly the one for 1935 to Carl von Ossietzky, the anti-militarist German journalist held by the Nazis in a concentration camp who had become an important international symbol for the struggle against Germany's rearmament. The prospect of a prize to Ossietzky led to the withdrawal from the committee of both Koht, at that time Norway's Foreign Minister, and Mowinckel, who served several times as Prime Minister. This was done to establish a separation between Norway as a state and the Norwegian Nobel Committee. At least Koht was also skeptical of the choice of Ossietzky as a Laureate. This was the first such withdrawal in the committee's history. The Storting then decided that no government minister could serve on the committee while in office. Hitler's reaction to the award was strong. He issued an order under which no German could receive any of the Nobel Prizes. (This affected two Chemistry Laureates, Richard Kuhn in 1938 and Adolf Friedrich Johann Butenandt in 1939 and Medicine Laureate Gerhard Domagk in 1939.) Ossietzky was not permitted to go to Oslo to receive the prize; he was transferred to a private sanatorium, but died 17 months later. The prize to Ossietzky illustrated how controversy could be combined with prestige for the prize, although this became much clearer over time than it was in 1936.

6. 1940–1945: The Second World War and Another Prize to the Red Cross

On April 9, 1940, Germany attacked Norway and two months later the entire country was occupied. The Norwegian government fled to London. Committee meetings were actually held during the first years of the war, but from 1943 with the committee members scattered, no further meetings were held. The early meetings focused on non-prize business. By underlining the Swedish nature of the Nobel Foundation, the Nobel Committee in Oslo escaped a German takeover of its Institute building.

No Peace Prize had been awarded for 1939 since the war had broken out well before the prize was normally announced. Later, during the war virtually no nominations came in. When the committee was able to meet again after the war, it decided to give the Peace Prize for 1944 to the International Committee of the Red Cross, the same Laureate as in 1917, with much the same reasoning. In the darkest hour the ICRC had "held aloft the fundamental conceptions of the solidarity of the human race." In so doing it had promoted the 'fraternity between nations' which Nobel had referred to in his will.

7. 1945–1966: The Cold War and the United Nations

In 1945, Norway joined the United Nations with considerable enthusiasm. There was little of the division and hesitancy that had characterized Norway's policy toward the League of Nations. The German attack on Norway had destroyed most of the earlier confidence in neutrality; so when the Cold War began and Norway felt it had to make a choice between East and West it definitely chose the West, first in the form of the Marshall Plan and then NATO. Norway became quite a loyal member of NATO, but remnants of the more traditional attitudes could be found in the policy of no foreign troops and no atomic weapons on Norwegian soil, in its negative attitude toward European and even Nordic integration and in a lingering skepticism toward Great Power politics and arms build-ups. The idealist component in Norwegian foreign policy now moved away from arbitration and mediation and more toward arms control and disarmament, aid to poor countries, and, increasingly, questions of human rights, certainly including those in Allied countries.

From 1945 to 1965 the Labor party dominated Norwegian politics. From 1949 to 1965 it also held a majority on the Nobel Committee, but the three Labor members rarely behaved as a group, since two were strongly Western-oriented (Martin Tranmæl and Aase Lionæs) and one was more neutral (Gustav Natvig Pedersen). The chairman of the committee from 1942, in effect from 1945 to 1966, Gunnar Jahn, was a stubborn Venstre politician; the Conservative leader C.J. Hambro was equally stubborn and had strong links back to the inter-war years. From 1949 to 1964 membership on the committee remained entirely unchanged. Again, tension within the committee was one strong factor behind the large number of years with no prize or postponed prizes (eight) during this period.

Of the 20 prizes awarded in this period, nine were in some way or other related to the United Nations, thus reflecting both the strong Norwegian support for the organization as such and the continuation of the long committee line going back to the Inter-Parliamentary Union and the League of Nations. Long-time US Secretary of State Cordell Hull was given the 1945 award primarily for his, and America's, strong leadership in the creation of the UN. In 1949 Lord John Boyd Orr of Brechin was honored as the founding director-general of the UN Food and Agricultural Organization, the first scientist to win the Peace Prize, not for his scientific discoveries as such, but for the way in which they were employed to "promote cooperation between nations." In 1950 the prize went to Ralph Bunche, the principal secretary of the UN Palestine Commission, for his mediation of the 1949 armistice between the warring parties. Bunche was also the first black person to receive the Nobel Peace Prize. More loosely connected to the UN, in 1951 veteran French and international labor leader Léon Jouhaux was the recipient of the Peace Prize. He had helped found the International Labor Organization in 1919 and had been active in the League of Nations. After the war he was a French delegate to the UN General Assembly. In 1954 the Office of the United Nations High Commissioner for Refugees, established in 1951, was honored, thereby underlining the long-standing interest of the Norwegian Nobel Committee in the question of refugees.

The Peace Prize in 1957 to Canada's Lester Bowles Pearson was given primarily for his role in trying to end the Suez conflict and to solve the Middle East question through the United Nations. As Foreign Minister of Canada he had become one of the leading UN statesmen of his period. In 1961, the prize was awarded to the second Secretary General of the UN, Dag Hammarskjöld, for strengthening the organization. Hammarskjöld is the only person to have received the prize posthumously, a few months after his death in a plane crash in the Congo; the Nobel statutes were later changed to make a posthumous prize virtually impossible. (In 1965 and 1966 a majority of the committee clearly favored giving the prize to the third Secretary General, U Thant, and even to the first, Norway's Trygve Lie, but chairman Jahn more or less vetoed this.) The United Nations Children's Fund (UNICEF), established by the UN General Assembly in 1946, was awarded the Peace Prize in 1965.

Most of the politicians who were given prizes related to the UN combined their UN work with a clear Western orientation in the Cold War. This went for Hull, Bunche, Jouhaux, and Pearson and, to a lesser extent, also Hammarskjöld. In this period no Communist politician was ever seriously considered for the prize. (Soviet diplomat and feminist Alexandra Kollontay was discussed in 1946–1947, but quickly rejected.) A whole series of Indians — Gandhi and Nehru, but also other politicians, philosophers and scholars — were considered, but all were found wanting in one way or another. Still the committee was reluctant to give the prize to politicians who were seen as too exclusively Western in their orientation. The only exception was George Catlett Marshall, Peace Laureate of 1953. Marshall's name was of course closely linked with the famous Marshall Plan, but the Cold War nature of his work was played down by the committee in favor of his role during the Second World War and his humanitarian work in general.

During this period too, the Norwegian Nobel Committee continued to honor individuals and organizations that had worked to strengthen the ethical underpinnings of peace. At least four of the awards fall under this category: the 1946 joint awards to Emily Greene Balch, co-founder and long-time leader of the Women's International League for Peace and Freedom and the acknowledged dean of the American peace movement, and to John Raleigh Mott, long-time executive of the Young Men's Christian Association (YMCA) and world ecumenical leader working for peace on the basis of the Bible. Thus, Balch followed closely in the footsteps of Jane Addams and Mott somewhat less closely than in those of Nathan Söderblom. In 1947 the prize went to two arms of the Quaker movement, the Friends Service Council in Britain and the American Friends Service Committee, for their work for social justice and peace, certainly including their relief work during and after the Second World War.

The humanitarian category of Peace Prize Laureates was well established through the prizes to Dunant, the two to the ICRC, to Nansen and to the Nansen Office. In this period it could be argued that at least in part, the prizes to Mott and to the Quakers and certainly the prize to the UN High Commissioner for Refugees fell in this category. Another clear-cut example was the prize for 1952 to Albert Schweitzer, the well-known medical missionary in Gabon who had started his work there as early as 1913. Schweitzer's ethical philosophy rested on the concept of 'reverence for life'. In the same category was the prize in 1958 to Georges Pire, Dominican priest and theologian, honored for his work on behalf of European refugees and even more for the spirit that animated his work. On the 100th anniversary of the founding of the Red Cross, the 1963 prize was divided between the Swiss International Committee of the Red Cross and the international League of Red Cross Societies, representing the two major arms of the Red Cross movement.

Neither was the disarmament category a new one. Possibly the prizes to Suttner and Arnoldson (who had favored an appeal stating, among other things, that "I want all armed forces to be abolished") and certainly to Henderson and Ossietzky could be seen as falling in this category. The continuity with the past was most clearly seen in the award in 1959 to Philip J. Noel-Baker. Noel-Baker had helped found both the League of Nations and the United Nations. His special interest was still disarmament and he had participated in the League's Conference on Disarmament in 1932. With the introduction of nuclear weapons, his work for disarmament became even more insistent. The biggest surprise in this category was the prize for 1962, awarded in 1963, to Linus Carl Pauling. Pauling had won the Nobel Chemistry Prize in 1954, but he then became increasingly preoccupied with the hazards of the nuclear arms race. He worked hard to bring about a test-ban treaty and the respect accorded him was strengthened by the signing of the partial test-ban treaty of 1963. Still, in many American circles Pauling was considered to harbor pro-Communist sympathies. The Western-oriented majority of the Norwegian Nobel Committee was actually against giving him the prize. What secured him the prize was chairman Jahn's threat to resign from the committee unless Pauling got it. Jahn, too, had become increasingly preoccupied with the danger of nuclear weapons.

One important category of Peace Prize Laureates was fully established in this period — those who worked for human rights. Some of the earlier Laureates had touched upon elements of human rights, although they had been primarily honored for other contributions. This went for Buisson, founder of the French League of the Rights of Man, Ossietzky, honored also for his right to speak out on the armament question, and Jouhaux, champion of economic and social rights. The first definite human rights prize was probably still the one for 1960 to Albert John Lutuli. The Zulu chief had been elected president-general of the African National Congress in 1952 and held this position until his death in 1967. He was thus in the very forefront of the struggle against apartheid in South Africa, a struggle which was receiving added international attention after the Sharpville massacre of March 1960. In a period when the ANC was about to change its tactics, Lutuli stood explicitly for non-violence. The Peace Prize to Lutuli is also often seen as signaling a change in the selection of Laureates in a more global direction. (More about this shortly.) In 1964 American civil rights leader Martin Luther King Jr. received the Peace Prize for his non-violent struggle against segregation, the American version of apartheid.

8. 1967–1989: The Cold War and the Globalization of the Prize

In the mid-1960s the membership of the Norwegian Nobel Committee changed. The Labor party lost its majority in 1965, and Jahn retired at the end of 1966. Labor held the chairmanship under Nils Langhelle (1967), Aase Lionæs (1968–1978), the first woman leader, and John Sanness (1979–1981). Lionæs had become a member of the committee as early as 1949; she was in fact the only woman on the committee until 1979. She was also the only one of the pre-1965 members continuing on the committee. Lionæs had tried to secure the Peace Prize for Eleanor Roosevelt, but failed; in general she did not particularly push female candidates. The non-Socialist majority held the chairmanship under conservative Bernt Ingvaldsen (1967) and Egil Aarvik (1983–1990) of the Christian People's Party, but it too rarely acted in unison. So, as in the earlier period, personal views were more important than party loyalties. In this period there were only three irregular prizes.

After 1965 political power fluctuated between Labor and non-socialist governments, but differences between the major parties were small on most foreign policy questions, with the primary exception of the very divisive issue of Norwegian membership in the European Community. In the Middle East, traditionally strong sympathies for Israel were increasingly balanced by a growing understanding of the Palestinian/Arab cause. Support for the UN remained very strong; the same was the case with backing for NATO, although the Vietnam war was to accelerate a more critical attitude to the United States, particularly among the youth and the increasingly important women groups. Impatience with the limited results achieved in arms control and disarmament, particularly on the nuclear side, was growing. On the Norwegian Nobel Committee this impatience was reflected in Chairman Aarvik's personal views. Norway's interest in human rights in most corners of the world was clearly also rising.

In this period four prizes were awarded to UN-related activities. In 1968, during the UN International Rights Year, and exactly twenty years after the approval by the UN General Assembly of the Declaration of Human Rights, René Cassin received the Peace Prize. Cassin was generally considered the father of the declaration, but had also served as vice-president and then as president of the European Court of Human Rights. (He had also been a French delegate to the League of Nations.) In 1969 the International Labor Organization (ILO) was honored. ILO was established in 1919 and it was the only organization associated with the League of Nations to outlive it; as a specialized agency of the UN, its work rested on the principle that peace had to be based on social justice. In 1981, on its thirtieth anniversary, the Office of the High Commissioner for Refugees received its second Peace Prize. Norway as a country had long made the largest per capita contribution of any country to this UN office. In 1988 the United Nations Peacekeeping Forces were honored. There was a strong feeling that as the Cold War was coming to an end, the UN ought to become more important and that this would be reflected in a new role for peacekeeping. In addition, the 1982 Peace Prize to Sweden's Alva Myrdal and Mexico's Alfonso García Robles could be considered at least in part a UN prize, since much of their disarmament work had been done in various UN negotiations.

Again no Communist politician was awarded the Peace Prize. Instead the human rights prizes to the Soviet dissident, and one-time creator of the Soviet hydrogen bomb, Andrei Dmitrievich Sakharov in 1975, to Polish labor leader Lech Wałesa in 1983, and to the 14th Dalai Lama in Tibet, Tenzin Gyatso, in 1989, the year of the Tiananmen Square massacre, were severely criticized by the Communist leadership in the three countries involved. Again the neutralist movement as such went unrecognized. On the Western side, German Chancellor Willy Brandt received the prize in 1971 for his *Ostpolitik*, an effort to bring East and West Germany, as well as Eastern and Western Europe, closer together. Brandt had spent the years from 1933 to 1945 in exile in Norway and Sweden, had excellent connections with Norwegian politicians and spoke perfect Norwegian. In 1974 former Japanese Prime Minister Eisaku Sato received the Peace Prize for his renunciation of the nuclear option for Japan and his efforts to further regional reconciliation. Sato was the first Asian to accept the Peace Prize, to the surprise of many in that part of the world, including even in Japan, who saw him as a rather conventional politician.

In 1973 the Nobel Peace Prize was awarded to US National Security Adviser and Secretary of State Henry A. Kissinger and North Vietnamese leader and negotiator Le Duc Tho for the 1973 Paris agreement intended to bring about a cease-fire in the Vietnam war and a withdrawal of the American forces. This award is definitely the most controversial one in the history of the Nobel Peace Prize. Le Duc Tho declined the Peace Prize, the only person to have done so, since there was still no peace agreement. Kissinger did not come to Oslo to receive the prize in person and soon indicated he wanted to return it, but was told the statutes did not permit this; two of the committee members resigned after it had become known that there had been disagreement and that they had in fact been against the award. (They supported Brazilian archbishop Helder Camara, who received a Norwegian people's prize instead.) Public reaction to the prize, both in Norway and internationally, was largely negative.

The 1973 controversy may have influenced the Storting to establish a new precedent under which the legislators themselves could no longer be members of the newly re-named Norwegian Nobel Committee. The members now tended to be either ex-politicians or persons not so explicitly connected with party politics. The most important reason behind the change, however, was a general desire to distinguish more clearly between the Storting itself and the non-parliamentary committees it appointed.

Regional crises represented nothing new in the Cold War. The Nobel Committee had previously awarded prizes to those who had worked to solve such crises, whether this be the crucial Franco–German conflict or the war between Paraguay and Bolivia. With the Cold War and the end of Western colonial rule over large parts of the world, such crises took on added prominence, also for the Nobel Committee. The situation in the Middle East was particularly difficult. In 1950 Ralph Bunche and in 1957 Lester Pearson had received the Peace Prize for their efforts there. In 1978, Egyptian President Mohamed Anwar al-Sadat and Israeli Prime Minister Menachem Begin were honored for the Camp David Agreement, which brought about a negotiated peace between Egypt and Israel. This agreement too, proved controversial. Only Begin came to Oslo to receive the award. A technicality prevented the American president, Jimmy Carter, from being the third Laureate; the committee actually wanted to include him, but he had not been nominated when the deadline expired on February 1 of that year.

In Western Europe the situation in Northern Ireland represented the bloodiest ethnic-national conflict. The Peace Prize for 1976 was awarded to Betty Williams and Mairead Corrigan for their efforts to end that conflict through a popular mobilization against violence. In Norway the Nobel Committee was strongly criticized for being late in recognizing the two women; they had in fact been given a Norwegian people's peace prize before the Nobel one. In 1987 Oscar Arias Sanchez, Costa Rica's president, was honored for his leadership in having the five presidents of Central America sign a peace agreement for the area. Both of these awards could be seen as the intervention of the Norwegian Nobel Committee in conflicts where progress toward peace had definitely been made, but the conflicts had been far from resolved. The committee clearly hoped that the prize itself would provide an added impetus for peace. This effect was very limited in Northern Ireland, but more significant in Central America, although it still took years before all the many conflicts there were more or less resolved.

On the issue of arms control and disarmament, referred to as 'the reduction of standing armies' in Nobel's will, the Nobel Committee, by general Western standards, again proved relatively radical. This was seen in the 1982 Peace Prize to Alva Myrdal and Alfonso García Robles, but even more clearly in the 1985 prize to International Physicians for the Prevention of Nuclear War (IPPNW). The committee had been so impressed by the cooperation between Soviet and American physicians within the IPPNW that it explicitly invited founders Evgeny Chazov and Bernard Lown to receive the award on behalf of the organization. Conservatives in West Germany, Britain, and the United States particularly criticized the committee's decision. (So did former committee chair Lionæs.)

In this, as in other periods, some humanitarians were also honored. Somewhat in the tradition of Boyd Orr, Norman Borlaug, an American of Norwegian descent, was selected in 1970 for his contributions to the 'green revolution' that was having such an impact on food production particularly in Asia and in Latin America. In 1979 Mother Teresa received the prize. She came from a family of Catholic Albanians, but lived most of her life in Calcutta, working for the poorest of the poor through her order, the Missionaries of Charity.

Among the more general peace advocates in this period, several have already been mentioned: Betty Williams and Mairead Corrigan, Alva Myrdal and the Dalai Lama. The best example was perhaps still the 1986 Laureate, Elie Wiesel. Wiesel was a Jewish survivor of the Holocaust and had become the leading interpreter of the relevance of this event for contemporary generations.

Human rights represented the fastest growing field of interest for the Norwegian Nobel Committee. The awards to the ILO and the Dalai Lama and, even more, to René Cassin, Andrei Sakharov and Lech Wałesa have already been mentioned. In 1974, Seán MacBride shared the prize with Eisaku Sato. MacBride had a multi-faceted background, but was honored primarily for his strong interest in human rights: piloting the European Convention on Human Rights through the Council of Europe, helping found and then lead Amnesty International and serving as secretary-general of the International Commission of Jurists. In 1977 the prize was awarded to Amnesty International itself. Founded in 1961, it was an increasingly important organization aimed particularly at protecting the human rights of prisoners of conscience. In 1980 the Argentinian human rights activist

Adolfo Pérez Esquivel was honored. Esquivel had founded non-violent human rights organizations to fight the military junta that was ruling his country. His message was also seen as relevant for much of the rest of Latin America. The apartheid regime in South Africa continued to preoccupy the Nobel Committee and the Norwegian public. In 1984 Bishop Desmond Mpilo Tutu was recognized for his non-violent struggle to bring apartheid to an end. The South African government strongly disliked the award, as it had Lutuli's, but again it let the Laureate travel to Oslo to receive it.

It was only in this period that the Nobel Peace Prize became truly global in its approach. The first Peace Prize to a person not from Europe and North America had been the one to Lamas in 1936. The next one was Lutuli's in 1960. Yet, even Lutuli's prize did not really signal an unmistakable trend, since only from the 1970s onwards did the Nobel Committee regularly award Asians (Le Duc Tho, Eisaku Sato, the Dalai Lama, in a sense also Mother Teresa), Africans (Anwar Sadat, Desmond Tutu) and Latin Americans (Adolfo Pérez Esquivel, Alfonso García Robles, Oscar Arias Sanchez). Thus, in the 1970s and 1980s there were as many Laureates from Africa, Asia, and Latin America combined as from North America and Western Europe combined. (In addition there were Andrei Sakharov and Lech Wałesa from Eastern Europe and Menachem Begin from Israel.)

One may ask why it took the Norwegian Nobel Committee so long to recognize persons from these other continents. The answer has several elements. For centuries Europe and North America dominated the rest of the world. There were few other independent actors. Reflecting this, very few nominations for the Nobel Peace Prize were submitted by persons from Asia, Africa and Latin America. In addition, most Western politicians simply did not pay much attention to what was going on in these vast regions; some even considered those who lived there inferior. Such feelings certainly affected Norwegians too, probably also some of the members of the Nobel Committee.

Mohandas Gandhi was, however, nominated five times and he was put on the committee's short list three times. In 1948 the committee awarded no prize; it indicated that it had found 'no suitable living candidate', a reference to Gandhi. It thus seems likely that he would have been awarded the prize if he had not been assassinated in January 1948. Still, the committee had had earlier opportunities to honor the man who, in hindsight, is generally seen as the leading spokesman of non-violence in the 20th century. Under the statutes then in force, Gandhi could have been awarded even the 1948 prize, as seen by the posthumous prize awarded to Hammarskjöld in 1961. Yet, a posthumous prize was an obvious complication. Gandhi had his supporters on the committee, but the majority felt that despite his own non-violence, violence had sometimes resulted from his actions, even before the bloody division between India and Pakistan; he was also perceived as too much of an Indian nationalist. Such feelings might have been affected by Norway's traditionally very close relationship to Britain, by a rapidly growing skepticism to neutrality in the Cold War and even by a more general underestimation of individuals from 'underdeveloped' parts of the world.

The reaction to apartheid in South Africa after the Sharpeville massacre was to modify such underestimation, but, as we have seen, this happened rather slowly. The decolonization process in Asia and Africa certainly also had an impact. All forms of racial stereotyping were banned from civilized public discourse. The growing emphasis on human rights furthered the globalization of the prize, as did the emphasis on finding a solution to regional crises in different parts of the world.

9. 1990– : Pluralist Globalization

Around 1990 huge changes were taking place internationally. The Cold War came to an end, with the collapse of the Soviet empire in 1989 and of the Soviet Union itself in 1991. Expectations were high for the new post-Cold War world, but it soon became obvious that an end to the Cold War did not signal the end of war and conflict. The arms race slowed down considerably, but it still continued in various parts of the world. Old conflicts lingered; many new ones arose. Human rights advanced greatly, with the emergence of new democracies in Central and Eastern Europe, in Latin America and Asia, and even in Africa, but almost half the world's population still lived under some form of dictatorship. The composition of the Norwegian Nobel Committee underwent few dramatic changes in the 1990s. The committee majority again moved left of center in terms of Norwegian politics, with the Labor party having two representatives and the Socialist Left one. After Aarvik's death in 1990, Labor's Gidske Anderson served as chair of the committee for only half a year, until illness forced her to step down. The committee chairman from 1991 to 1999, Francis Sejersted, was a Conservative professor of history. In 2000 former Labor cabinet minister Gunnar Berge became the new chairman. From 1979 the committee regularly had two women members; from 2000 it even had a female majority. In the 1990s the prize was awarded on a regular basis every year.

The Norwegian Nobel Committee celebrated the end of the Cold War with the 1990 Peace Prize to Mikhail Sergeyevich Gorbachev, President of the Soviet Union, the person who, in the Committee's opinion, had done more than any one else to bring the Cold War to an end. Encouraged by the end of the Cold War, the committee was also prepared to intervene even more frequently than before in regional conflicts around the world in the hope that the Nobel Peace Prize could not only award deeds done, but also provide an added incentive for peace. The prize in 1993 to Nelson Mandela and Frederik Willem de Klerk could be regarded as a success in that respect, although it came at a stage when most of the transition from apartheid to democracy had already been accomplished.

In 1994, the Peace Prize was awarded to Palestinian leader Yasser Arafat, Israeli Prime Minister Yitzhak Rabin and Israeli Foreign Minister Shimon Peres for the Oslo Agreement, which brought about a mutual recognition and a framework for peace between the Palestine Liberation Organization (PLO) and Israel. The three politicians had accomplished much, but they were still far from establishing a final peace between Israelis and Palestinians. The award resulted in one member leaving the committee, the leading spokesman in Norway for the Likud party in Israel. This was the third resignation in the history of the Norwegian Nobel Committee. In 1996, the prize was awarded to East Timorese leaders Bishop Carlos Filipe Ximenes Belo and José Ramos-Horta. The tragic situation in East Timor after the Indonesian invasion in 1975 had been almost forgotten internationally. Due to the effect of the Nobel Peace Prize and, even more, of the Indonesian economic and political collapse in 1997–1998, East Timor was able to start on the road toward independence. In 1998 the committee honored Northern Irish leaders John Hume and David Trimble. Through the Good Friday Agreement of that year, the major parties to that protracted conflict agreed on the principles for its resolution, although it might take years before the agreement is fully implemented. In 2000 the Peace Prize was awarded to South Korean President Kim Dae Jung, both for his 'sunshine policy' of contacts and cooperation with North Korea and his long-standing commitment to human rights in South Korea and elsewhere.

The Norwegian Nobel Committee also further strengthened its somewhat radical profile within the field of arms control and disarmament. Two such prizes were awarded in the 1990s. The first one came in 1995, on the 50th anniversary of the atomic bombs over Hiroshima and Nagasaki, to Joseph Rotblat and the Pugwash Conferences on Science and World Affairs. Rotblat had initially worked on the Manhattan Project, which created the bombs, but had left the project to take up a life-long struggle against nuclear weapons. He had helped create the Pugwash Conferences where since 1957, scientists from the United States, the Soviet Union and many other countries had met in an effort to reduce the role of nuclear weapons in international relations. The second prize came in 1997 when the International Campaign to Ban Landmines (ICBL) and its coordinator, Jody Williams, were honored for their work to ban and remove anti-personnel land mines and to support the victims of such mines.

In the 1990s the human rights tradition was extended by prizes to two women. In 1991 the Peace Prize was awarded to Aung San Suu Kyi, the leader of the opposition against the Burmese military regime. Her party won an overwhelming victory in the 1990 election, but she was then confined to house arrest. While her cause now came to receive broad international support, the military regime continued in power. Somewhat more controversial was the 1992 award, on the 500th anniversary of Columbus's discovery of America, to Rigoberta Menchú Tum, the Maya Indian campaigner for human, particularly indigenous, rights in Guatemala and the rest of Latin America. The humanitarian tradition was continued through the 1999 award to Médecins Sans Frontières (MSF) — or Doctors Without Borders — for its "pioneering humanitarian efforts on several continents." The work of MSF clearly had a human rights dimension in addition to the humanitarian one. As already mentioned, the 2000 award to Kim Dae Jung also combined two traditional elements in the history of the Peace Prize.

10. The Nobel Peace Prize through 100 Years: Some Conclusions

Thus, some lines of development can be distinguished in the almost 100 year history of the Nobel Peace Prize. First, although the Norwegian Nobel Committee never formally defined 'peace', in practice it came to interpret the term ever more broadly. This approach could have its pitfalls, but avoided the danger of locking the committee into fixed categories and gave the committee flexibility to adapt to new concerns. In the early years, the emphasis was definitely on the organized peace movement and the codification of international law, but even in the very first year of the Peace Prize the first humanitarian, and five years later, the first statesman were selected. Later the balance shifted away from the organized peace movement and international jurists, although some of them continued to be selected and the category came to include church leaders and even a

Holocaust interpreter. Humanitarians became more numerous, and this category came to include scientists who worked to alleviate hunger. Disarmers became more numerous too, and this category came to include those who supported limited arms control and not necessarily full disarmament. Different kinds of statesmen were awarded the Peace Prize, some for addressing global concerns, others for helping to solve regional crises, still others for the general principles they espoused. The human rights category was added to the list and gradually became perhaps the most numerous one.

Second, from a slow start, the list of Laureates became increasingly global, so that by the 1970s all continents except Australia and Oceania were represented. In the nominations and correspondence to the committee, it is easy to see how a prize to one continent stimulated interest in the prize in this area. Third, although Bertha von Suttner was awarded the Peace Prize in 1905, particularly in the early decades few women were selected. In recent decades, this too has changed, although not as dramatically as the geographical distribution of the Laureates, so that by 2000 ten women have received the Nobel Peace Prize. Fourth, the Norwegian Nobel Committee has increasingly come to use the Peace Prize not only as a reward for achievements accomplished, but also as an incentive for the Laureates to achieve even more. This may be said to reflect the growing courage of the committee members or, perhaps more accurately, the increasing stature of the Nobel Peace Prize.

No prize will be able to establish a 'perfect' historical record, whatever that might be. Most observers will agree that the omission of Gandhi from the list of Nobel Laureates is a serious one, but it might be the only one of such a nature. There may well have been some Laureates that perhaps should not have received the prize, but still did. But there is not much of a consensus on which ones these Laureates are. Controversy is certainly no good judge in this respect. (These days, when even Mother Teresa is considered controversial by some, it may also be difficult to know what is controversial.) In historical hindsight, several of the more controversial prizes are now considered among the most successful ones (Ossietzky, Lutuli, Sakharov, the Dalai Lama, Gorbachev.) On the other hand, the prize to Kissinger and Le Duc Tho shows that controversy is no guarantee of historical success. On the whole, however, after taking into consideration what a treacherous field 'peace' is and also the record of the many other peace prizes, it can certainly be argued that the standing of the Nobel Peace Prize would not have been what it is if it had not been for its highly respectable record.

This essay has attempted to place the history of the prize within a Norwegian context. This is natural since the committee members through these 100 years have all been Norwegians. Until 1936 they sometimes included even prominent members of the Norwegian government; until the 1970s they were frequently members of the Storting. Later they were often ex-politicians, many of them having served in prominent positions. Some of the politicians honored, from Roosevelt to Arafat, Peres and Rabin, may well have served Norwegian state interests in the sense that their selection fitted well into government policy. On the more speculative side, the non-award to Gandhi may also have been influenced by Norway's close relationship to Britain, and after the Second World War any award to the leading figures behind the movement toward European economic and political integration was clearly difficult in a country as divided on that issue as was Norway. (Brandt was only a partial exception.) On the other hand, some of the committee's selections were clearly problematic from the point of view of the Norwegian government. The best illustrations of this were probably the awards to Ossietzky and the Dalai Lama.

In principle almost everyone would prefer a Nobel Committee with an international membership. In practice, however, an international committee would have faced serious problems. (What would such a committee have done during the Cold War?) The connections to Norwegian values, as well as to Norwegian politics, may be regarded as questionable for the prestige of the Peace Prize, but it may in fact have had its advantages. Thus, after the Second World War hardly any term has been and still is more popular in Norwegian foreign policy parlance than 'bridge-building'. While an increasingly rich Northern state firmly attached to the West and with strong sympathies for Israel, Norway has been concerned with building bridges to the East, to the South, and increasingly to the Palestinians and other Arabs. It is a separate question how realistic such attitudes are as a basis for a country's foreign policy, but as a basis for prize selections, a blend of idealism and realism may not be so bad.

The values that underpinned the Nobel Peace Prize were concretely defined by Norwegians, but they were part of a wider Scandinavian and Western context. They represented the Norwegian version of Western liberal internationalism. Thus, the Norwegian Nobel Committee has been a strong believer in international organizations, from the Inter-Parliamentary Union to the League of Nations and the United Nations. Organizations and rules had been employed to contain conflicts within Norway; they could also temper international strife. Small nations almost instinctively prefer international law to the might they do not possess, and they believe in the arbitration, mediation and peaceful solution of international disputes. In a similar way, the Nobel Committee believed in humanitarian assistance to the weak and the poor, in arms control and disarmament, and, more and more fervently, in human rights generally.

When we look at the nationalities of the Laureates, we also get an idea of where liberal internationalism has been most strongly represented (or has been perceived by the Norwegian Nobel Committee as being most strongly represented). Virtually all of the organizations honored had clear roots in this Western ideology. Although liberal internationalism was in many ways ideally suited for smaller powers, it also had many supporters in the Great Powers. On the individual side, nineteen of the Laureates have come from the United States, representing both leading politicians — two presidents, one vice-president, five secretaries of state — and those more distant from and skeptical to the centers of power (Addams, Balch, Pauling, King, Williams); twelve have come from Great Britain, again reflecting both traditions, Austen Chamberlain and Joseph Rotblat perhaps representing the extremes; eight have been French, four have been German. Five have been Swedish (Arnoldson, Branting, Söderblom, Hammarskjöld and Myrdal). Two have been Norwegian (Lange, Nansen).

Thus, perhaps, in compiling its record through these 100 years, the Norwegian Nobel Committee has actually been able to be both very Norwegian and quite international at the same time.

Peace 1901

Jean Henry Dunant (1828–1910)

Founder of the International Committee of the Red Cross, Geneva, Switzerland

Peace 1905

Baroness Bertha Sophie Felicita von Suttner,
née Countess Kinsky von Chinic und Tettau
(1843–1914)

Honorary President of Permanent International Peace Bureau, Berne, Switzerland and Author of 'Lay Down Your Arms'

Peace 1964

Martin Luther King, Jr. (1929–1968)

Leader of Southern Christian Leadership Conference, USA

Peace 1979

Mother Teresa (1910–1997)

Leader of Missionaries of Charity, Calcutta, India

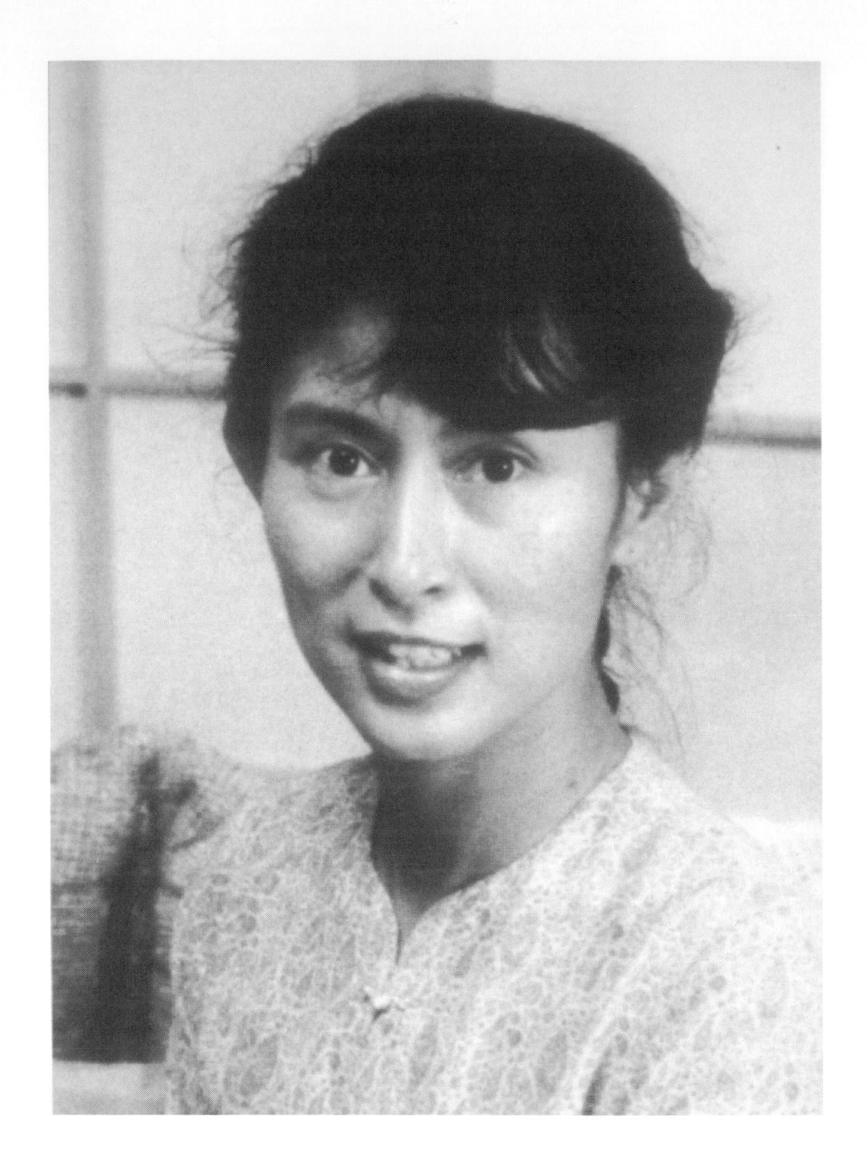

Peace 1991

Aung San Suu Kyi (1945–)

Peace 1993

Nelson Mandela (1918–)

ECONOMIC SCIENCES

Sveriges Riksbank (Bank of Sweden) at their tercentenary in 1968 instituted *The Bank of Sweden Prize in Economic Sciences in Memory of Alfred Nobel* and placed an annual amount at the disposal of the Nobel Foundation as a basis for a prize to be awarded by the Royal Swedish Academy of Sciences. Nobel Prize rules will, mutatis mutandis, be followed regarding nomination of candidates, prize adjudication, prize award and prize presentation.

The Sveriges Riksbank (Bank of Sweden) Prize in Economic Sciences in Memory of Alfred Nobel 1969–2000

*Assar Lindbeck**

Introduction

In conjunction with its tercentenary celebrations in 1968, Sveriges Riksbank (Bank of Sweden) instituted a new award, 'The Central Bank of Sweden Prize in Economic Sciences in Memory of Alfred Nobel', on the basis of an economic commitment by the bank in perpetuity. The award is given by the Royal Swedish Academy of Sciences according to the same principles as for the Nobel Prizes that have been awarded since 1901.

The procedures for selecting the laureates are also the same. Each year the Academy receives some 250 nominations, usually covering a little more than one hundred nominees. (Unsolicited suggestions from persons who have not been asked to submit nominations are not considered.) The Economics Prize Selection Committee of the Academy (with five to eight members) commissions expert studies of the most prominent candidates, sometimes by Swedish experts but usually by experts from other countries who are reputed internationally. The Prize Committee presents its award proposal to the Social Science Class of the Academy in the form of a report, with an extensive survey of the main candidates who are considered for a Prize. The report presents arguments in support of the proposal and includes all the solicited expert studies. Finally the entire Academy meets to take the final award decision, usually in October.

*Professor of International Economics at the Institute of International Economic Studies, Stockholm University, and IUI Stockholm, Sweden.

What criteria have guided the awards so far? And what have been the main problems when selecting the Laureates?

It is useful to start a discussion of these issues with a rough classification of the various types of Economics Prize awards given so far. It should be kept in mind, however, that all such classifications are rather arbitrary since the multidimensional nature of scientific contributions makes it difficult to avoid overlaps.

1. A Classification of Prizes for the First 32 Years

1.1. *General Equilibrium Theory*

Obvious examples of this type of award are the prizes to Paul Samuelson (1970) for having "developed static and dynamic economic theory;" to Kenneth Arrow and John Hicks (1972) for "their pioneering contributions to general economic equilibrium theory and welfare theory;" to Gerard Debreu (1983) for "his rigorous reformulation of the theory of general equilibrium;" and to Maurice Allais (1988) "for his pioneering contributions to the theory of markets and efficient utilization of resources." (See the table at the end of the article for an attempt to classify the awards into various fields of research.)

Contributions in this category have dealt largely with the analytical structures of theoretical economic models, often highlighting the formal similarity of these structures, and clarifying the conditions for consistency, equilibrium, stability and efficiency of the economic system. Often, these contributions also have included important comparative static experiments, i.e. analyses of how equilibrium positions change in response to changes in various exogenous factors (parameters).

It is largely due to the above-mentioned theorists that general equilibrium theory has become the basic approach in theoretical economic analysis. For instance, Hicks formulated conditions for multimarket stability and extended the applicability of the static method of analysis to several periods. He also initiated rigorous dynamic analysis of capital accumulation. Because it was deeply anchored in microeconomic theories of the behavior of individual consumers and firms, the models developed by Hicks offered far better ways to study the consequences of changes in various parameters than did earlier general equilibrium models. Hicks also presented a celebrated aggregate general equilibrium model with four markets — commodities, labor, credit and money — the so-called IS–LM model.

Samuelson's work was not only a continuation of the contributions by Hicks; it also represented a discontinuity, i.e. a breakthrough, in terms of analytical sophistication. This is recognized in the prize citation, which declares that Samuelson "actively contributed to raising the level of analysis in economic science." It is hardly an exaggeration to say that he single-handedly rewrote considerable parts of central economic theory: microeconomic theory, static and dynamic, partial and general equilibrium theory, as well as welfare economics. By extracting interesting inferences from simple mathematically formulated models, exploiting effectively the second-order conditions of maximization procedures, he derived results which today still rank among the classical theorems of economics.

Arrow's and Debreu's main contributions to general equilibrium theory were to achieve greater generality by applying more powerful mathematical methods, such as the theory of convex sets. The generality allowed them to define the concept of a good so broadly, that the same theory may be used not only in static equilibrium analysis, but also in the analysis of the spatial distribution of production and consumption activities, intertemporal analysis and the analysis of decision-making under uncertainty. Arrow also highlighted the difficulties of deriving social welfare functions from individual preferences — Arrow's so called 'impossibility theorem'.

Maurice Allais' contributions, made largely in the 1940s, have great similarities both with Paul Samuelson's (contemporaneous) work and Arrow's and Debreu's (later) contributions. A special feature of Allais' work is that he describes the economy's path to equilibrium as a process by which competition removes all 'surpluses' in firms. Allais' analysis covers the case where returns to scale in production give rise to natural monopolies. His contributions thereby laid the foundation for a school of Post-War French economists who analyzed the conditions for an efficient use of resources in large public monopolies (such as Electricité de France and SNCF, the state railway system). Allais also anticipated parts of the modern theory of economic growth.

1.2. *Macroeconomics*

Numerous prizes have been given to macroeconomics, i.e. that branch of economic analysis that explains the behavior of the national economy as a whole in terms of a number of broad aggregates, such as private consumption, investment, exports, imports, government spending of goods and services, etc. Some of the awarded contributions in this field concern sectors ('submodels') of national economies, while others deal with an entire national economy.

An award in macroeceonomics that refers both to special sectors and to the entire national economy is the 1976 Prize to Milton Friedman. The prize citation referred to his contributions to "consumption analysis, monetary history and theory." Milton Friedman's book, *A Theory of the Consumption Function* in 1957, is a successful attempt to combine formal theory and its empirical application for a specific sector of the economy. His extensive empirical study of the monetary history of the United States (together with Anna Schwartz) may be regarded as an example of rather 'pure' empirical research, even though the study clearly was based on a theoretical framework emphasizing a monetary interpretation of macroeconomic fluctuations.

Franco Modigliani (awarded the Prize in 1985) developed two important building blocks in macroeconomic models, namely submodels of private consumption and the financial sector. In particular, in his life-cycle theory of saving, Modigliani studied the consequences for household saving of changes in demography and economic growth. Together with Merton Miller he also laid the foundation for the field 'corporate finance'. The Modigliani–Miller theorem states the conditions under which the value of a firm in the stock market is influenced (or not influenced) by the dividend policy of the firm, and the way the firm finances its investment, e.g. via equity capital or borrowing.

The Prize to James Tobin (1981) is another example of an award for theoretical contributions concerning specific sectors of a national economy — the award being given for his analysis of "financial markets and their relation to expenditure decisions, employment, production and prices." Tobin's way of modeling interactions between financial and real sectors quickly became an integrated part of macroeconomic models for national economies, with an important role played by the relation between the market value of a capital asset and its reproduction costs, the so-called 'Tobin's q'. Adding the stock of real assets — land, buildings, inventories and claims on raw materials — Tobin's portfolio model also becomes the natural analytical tool with which to analyze direct effects on product prices of changes in the supply of money.

Lawrence Klein (awarded the Prize in 1980) also made important contributions to macroeconomic research. The prize citation emphasized "the creation of econometric models and their application to the analysis of economic fluctuations and economic policies." One of Klein's main achievements was to analyze the effects of economic policies by way of statistical model simulation. He also made important contributions in developing forecasting techniques. His analysis originally ran in the framework of Keynesian–type macrotheories, but his models tended to become more eclectic over time. They also became more and more detailed, ultimately covering more than one hundred estimated equations.

Robert Lucas, awarded the Prize in 1995, has also furthered macro-economic model building in a fundamental way. In particular, he has emphasized the role of expectations in macroeconomic analysis. He is particularly renowned for developing the consequences of 'rational expectations' among economic agents, according to which these exploit all available information and do not make *systematic* expectational mistakes. Lucas also analyzed the consequences for the macroeconomy of changes in the 'economic policy regime', i.e. the way government and central bank policies respond to changes in the economy. In particular, he has shown how conventionally statistically estimated macroeconomic behavior functions for the private sector may become unreliable after a change in the policy regime — the so-called 'Lucas Critique' of traditional macroeconometric estimations. He has also suggested ways of avoiding this problem.

The shared Prize to James Meade and Bertil Ohlin (1977) for their contribution to "the theory of international trade and international capital movement" is another example of a contribution concerning a specific sector of a national economy: the sector of foreign transactions. In the case of Ohlin, the award referred to his development of a theory of international and interregional trade, designed to explain both the causes and the consequences of trade — known as the Heckscher–Ohlin model. Ohlin showed that the trade patterns of individual countries depend on their proportions of available factors of production (capital and labor), and that international trade tends to equalize the returns to these factors among countries. James Meade analyzed trade policy in a world with various market distortions, hence anticipating the theory of 'second best' allocations of resources. He was also a pioneer in the field for the theory of open–economy macroeconomics. Of particular importance was Meade's analysis of the relation between internal and external balance, and the relation between targets and instruments of economic policy.

However, the foundations for today's theory of open-economy macroeconomics were largely constructed by Robert Mundell (awarded in 1999) — the so-called Mundell–Fleming model. In particular, he developed a powerful yet simple model for the analysis of monetary and fiscal policy, i.e. stabilization policy, in open economies. Specifically, Mundell introduced foreign trade and capital movements into Hick's IS–LM macroeconomic model for a closed economy. He showed that the effects of stabilization policy hinge crucially on the degree of international capital mobility. Mundell demonstrated the far-reaching importance of the exchange rate regime: under a floating exchange rate, monetary policy is powerful while fiscal policy tends to be rather powerless, whereas the opposite is true under a fixed exchange rate. Mundell is also a pioneer in the analysis of optimum currency areas, which deals with the advantages and disadvantages for countries in relinquishing their monetary sovereignty in favor of a common currency.

While the awards to macroeconomics discussed above referred to contributions concerning short-term macroeconomic fluctuations, Robert Solow was rewarded (in 1987) for his contributions to the theory of long-term macroeconomic growth. His main contribution was to build a mathematical model (in the form of a simple differential equation) describing how the process of capital accumulation generates rising productivity. The capital intensity of production — the volume of capital per worker — is determined by the prices of capital and labor. Due to diminishing return to capital, the economy in this model will, in the long run, approach a situation where productivity growth is driven only by technological progress. Solow also developed a model of economic growth in which new technology was embedded in newly produced capital goods, the so-called 'vintage model' of economic growth. Based on his theoretical models, Solow also pioneered in empirical research on the determinants of economic growth — the so-called 'growth accounting'.

The shared prize to Arthur Lewis and Theodore Schultz (in 1979) also referred to economic growth, though at a less abstract level than the work of Solow. The prize citation referred to their research on "economic development with particular consideration of the problems of developing countries." The award to Lewis recognized particularly his two long-term growth models for less developed countries — emphasizing the consequences for economic growth of an elastic supply of labor, and the determinants of the terms of trade for countries that export tropical products. The award to Schultz honored his analysis of the role of investment in human capital for economic development, particularly in agriculture. Both Lewis and Schultz were concerned with combining their theoretical reasoning with empirical data, though they used the traditional expository techniques of economic history rather than formalized statistical or econometric testing techniques. Schultz emphasized the apparent efficiency in the agricultural sector in less developed countries, considering existing constraints with respect to resources and knowledge available in these countries. Lewis instead focused on the tensions between a large and stagnant agricultural sector, with a low marginal product of labor, and a dynamic industrial ('capitalist') sector, which sometimes is in the nature of an economic enclave.

1.3. *Microeconomics*

A number of awards have also been given for contributions in microeconomic theory, dealing with decision-making by individual households and firms, and the allocation of resources among different uses and production sectors in the economy. One example is the Prize to George Stigler [1] for his studies of "industrial structures, functioning of markets and causes and effects of public regulation." He also analyzed how economic regulations, in fact, are enforced by politicians and public-sector administrators. He showed, for instance, that regulators often become dominated by those that are supposed to be regulated — so called 'regulatory capture'. In a similar vein as Friedman, Stigler represents a pronounced positivist tradition, emphasizing analytical simplicity and the importance of empirical application.

Stigler was also one of the pioneers in the field of 'information economics', introducing information costs explicitly in his analysis. Other prizes have also been given to this field. James Mirrlees and William Vickrey (awarded the Prize in 1996) made pioneering work about the consequences of various limitations in the information of individuals, including 'information asymmetries' among economic agents. In particular, both studied incentive problems in connection with asymmetric information. It turns out that such information asymmetries are of great importance for the functioning of markets such as insurance and credit markets. Mirrlees did fundamental work on the consequences for taxation of asymmetric information between the government and private agents. Vickrey clarified the properties of various types of auctions. His insights have been crucial for developing efficiently functioning auctions of rights to broadcast, landing permits at airports, television rights as well as sales of government assets ('privatization').

Though financial economics relies on analytical techniques similar to those of traditional microeconomics, over time it has become a field of its own, with an enormous expansion during the last two decades. Tobin and Modigliani were mentioned above as early contributors as parts of their construction of important building blocks to macroeconomic theory. However, the field of financial economics is today built mainly on foundations laid in the 1950s and 1960s by Harry Markowitz, Merton

Miller and William Sharpe (jointly awarded the Prize in 1990). While Markowitz' contribution was to construct a microtheory of portfolio management of individual wealth holders, Merton and Sharpe developed equilibrium analysis in financial markets. More specifically, Sharpe developed a general theory for the pricing of financial assets. Miller made important contributions in the field of corporate finance (to begin with, partly in cooperation with Frances Modigliani). In particular, Miller clarified which factors determine share prices and capital costs of firms.

Subsequently, Robert Merton and Myron Scholes were given the Prize (in 1997) for their analysis of price formation of so-called derivative instruments such as options, which are claims on underlying financial instruments including shares and foreign exchange. (The late Fisher Black was also instrumental for this achievement.) These contributions were a necessary condition for the subsequent development of today's huge markets for various types of derivative instruments. These markets have increased the possibility for individual agents to choose adequate risk levels according to their own preference, regardless of whether they choose low or high exposure to risk.

1.4. *Interdisciplinary Research*

Several prizes have also been awarded to economists who have widened the domain of economic analysis to new areas. James Buchanan got his Prize (in 1986) for his research on the boundary between economics and political science, or more specifically, "for his development of the contractual and constitutional basis for the theory of economic and political decision-making." This research made him one of the founding fathers of the 'public choice' school, which analyzes the driving forces behind political decisions and tries to endogenize political behavior in models of national economies. Rather than looking at politicians as individuals that are supposed to take care of the 'general good' in society, the public choice school assumes that politicians are motivated by considerations similar to those explaining the behavior of other agents, including striving for personal benefit and a desire for power.

Gary Becker (awarded the Prize in 1992) has instead worked on the borderline between economics and sociology, in particular for his research about the family. He has not only analyzed the 'economic' behavior of families — labor supply, consumption, household production and household saving — but also behavior that has not earlier been much considered by economists, such as education, marriage, childbirth, and divorce. He has shown how both economic considerations influence choice in these areas, and analyzed 'social interaction' between individuals outside the market system, reflected in the prize citation: "for having extended the domain of microeconomic analysis to a wide range of human behavior and interaction, including nonmarket behavior." Becker's influence today extends far outside economics, in particular to the so-called 'rational choice' school in sociology.

Ronald Coase (awarded the Prize in 1991) has instead made important contributions on the borderline between economics, law and organization. In particular, he showed which factors determine the size of firms. He also clarified the condition under which voluntary contracts between private agents can resolve problems with 'external effects' of production, an important example being pollution. These contributions are reflected in the prize citation: "for his discovery and clarification of the significance of transaction costs and property rights for the institutional structure and functioning of the economy." Coase's concept of transaction costs has become an important foundation for the theory of contracts and for the whole field of 'law and economics'.

The Prize to Herbert Simon (in 1978) may also be regarded as an interdiciplinary award. The prize citation referred to his research on "the decision-making process within economic organizations." In particular, Simon challenged some basic building blocks of microeconomic theory, in particular, the maximization principle and the assumption about full ('unbounded') rationality. On the basis of both empirical evidence and psychological theory, Simon argues that decision-makers usually do not try to choose a 'best' alternative, as assumed in traditional microeconomic theory, but that they are content with a 'satisfactory' outcome, i.e. they try to find acceptable solutions to acute problems. This has made Herbert Simon a main contributor in the field of administrative (management) science.

Simon Kuznets (1971) has instead done empirical research on the borderline between economics and history, reflected in the prize citation "for his empirically founded interpretation of economic growth." This Prize is an example of an award for inductive rather than deductive analysis. Kuznets' ambition was to make empirical generalizations from data interpreted with a minimum of formal models and without relying on complex statistical techniques. Important examples include the celebrated 'Kuznets' curve' of the U–shaped relation between GDP and income inequality, as well as his findings that the long-run average propensity to consume out of income tends to be constant in time-series data, whereas it tends to fall in cross-section data. More generally, Kuznets has exploited data for very long periods of time to extract regularities, in particular, by characterizing economic growth and the distribution of income in different nations at different times.

The Prize to Robert Fogel and Douglass North (in 1993) is another award on the boundary between economics and history. The Academy cited them "for having renewed research in economic history by applying economic theory and quantitative methods in order to explain economic and institutional change." Fogel's main contributions have been to clarify the role of the railways for the development of the national economy in the United States, and the economic role of slavery. By comparing the factual development with a counterfactual benchmark, Fogel concluded that previous studies of economic growth in the United States had vastly overestimated the importance of railways. He also concluded that slavery was not abolished because of falling profitability of the slave system, but rather because of humanitarian considerations. Douglass North has shed new light on the economic development in Europe and the United States before and in connection with the industrial revolution, including the roles of sea transport and changes in the pattern of regional specialization and interregional trade. He has also been a pioneer in analyzing the role of institutions, such as property rights, for economic development, as well as the importance of different types of transaction costs. In these fields he developed and applied the ideas initially launched by Ronald Coase.

Research on the borderline between economics and philosophy was honored with the prize to Amartya Sen (in 1998) "for his contributions to welfare economics." Sen scrutinized the philosophical foundations of collective decisions and welfare evaluations, including problems of evaluating the distribution of income and wealth. He has also constructed influential indices to measure income distribution and poverty. Sen has also analyzed the determinants and consequences of starvation in a number of less developed countries. These empirical studies of actual famines show that reduced aggregate supply of food has not always been the most important factor for starvation catastrophes which, in some cases, have instead been caused by redistribution of income to the disadvantage of the poor.

The award to Friedrich von Hayek and Gunnar Myrdal (in 1974), too, had a strong interdisciplinary flavor. While their early contributions on business cycles and monetary phenomena in the 1930s comprised quite abstract (though non-mathematical) economic–theory structures, their works from the early 1940s instead dealt with the interrelations between economic, social and political processes. Hayek is perhaps known among economists mainly for emphasizing the information and incentive content of the price system. However, he has given particular attention to the importance for individual behavior of the institutional framework for economic decisions, including the political constitution and the legal rules that define contracts and property rights. In these fields, Hayek's work parallels the work by Buchanan and Coase. Hayek has also emphasized the importance of 'spontaneous' social order in contrast to planned institutional designs.

Gunnar Myrdal has combined economic analysis with a broad sociological perspective in order to show how social, economic and political forces interact, often generating vicious or virtuous circles. In fact, Myrdal has described his methods of analysis of 'mutual causation' as a generalization of Knut Wicksell's 'cumulative process' in monetary theory. The most important example is Myrdal's study of the 'Negro Problem' in the United States in his book *An American Dilemma* (1944). This work not only influenced social science research, it also played an important part in the political discussion on segregation and integration of ethnic groups in various countries. The Supreme Court in the United States referred to Myrdal's book when outlawing segregation. Myrdal applied a similar approach with 'mutual causation' in his subsequent work on poverty and economic development in South Asia.

1.5. *New Methods of Economic Analysis*

Though several of the awards discussed above could perhaps be regarded as 'method awards', there are more clear-cut examples. One case in point is the joint Prize to Ragnar Frisch and Jan Tinbergen (the very first award in 1969) for their pioneering work on econometric model building, i.e. the integration of economic theory and statistical methods. The prize citation was "for having developed and applied dynamic models for the analysis of economic processes." While Frisch developed general methods of dynamic and econometric analysis, Tinbergen pioneered in applying such methods empirically. Tinbergen's main achievement was to make rigorous statistical tests of the realism of alternative business cycle theories. Frisch and Tinbergen were also instrumental in developing a formalized theory of the relation between instruments and targets of economic policy — a contribution paralleling Meade's analysis of similar issues. Frisch and Tinbergen gave these theories a form that was favorable for empirical quantification and statistical testing. Frisch based his analysis partly on a system of national accounts for Norway, the so-called 'oekosirk system' (income and expenditure flows), while Tinbergen pursued much of his empirical policy analysis in the context of econometric macro models for the Netherlands.

The Prize to Frisch's countryman Trygve Haavelmo (in 1989) honored further development of Frisch's work. More precisely Trygve Haavelmo was awarded "for his clarification of the probability theory foundations of econometrics and his analysis of simultaneous economic structures." Haavelmo showed how methodology of mathematical statistics could be applied to draw stringent conclusions about complex economic relations from a random sample of empirical observations. These methods could then be used to estimate relations derived from economic theories and to test these theories. He also showed that misleading interpretations of partial relations between economic variables due to interdependencies can be avoided if these relations are estimated simultaneously.

Another important breakthrough in econometrics was achieved by James Heckman's and Daniel McFadden' theories and methods for empirical analysis of individual behavior — microeconometrics. Their awarded contributions (in 2000) have greatly improved the possibilities to analyze data about large groups of individual agents, households as well as firms.

Heckman has developed methods to avoid biased statistical estimates in situations when the analyzed sample of data is no-random — the well-known Heckman correction (the Heckit method). Such situations often occur when only some agents, often with characteristics that are unobservable to the researcher, do not appear in the sample. Important examples are studies of wage formation and the return on education, since individuals who do not work and do not have the type of education that is studied are often not included in the sample.

McFadden has developed methods to analyze the choice by individual agents among a limited (finite) number of alternatives, so-called discrete choice. Important examples are the choice of profession, occupation, residence and means of transportation. A seminal contribution by McFadden is his so-called conditional logit analysis. The method utilizes not only observable facts about characteristics associated with individual agents and information about each available alternative choice. Unobservable differences among individuals and among alternatives are also exploited; they are represented by random error terms.

Another example of an award for important methodological developments is the Prize to Wassily Leontief (in 1973) "for the development of the input-output method." This methodology highlights the interdependencies between different sectors of the economy in quantitative form. The analysis is also well suited to an analysis of the short-term effects of shocks in one sector on other sectors of the economic system. The candidacy of Leontief was greatly enhanced by the fact that he also pioneered in applying his method to empirical data. There is a parallel between Tinbergen's contribution to make macroeconomic theory empirically operational and Leontief's inter-industry analysis.

The Prize to Richard Stone (in 1984) for "having made fundamental contributions to the development of systems of national accounts" similarly awarded important new methods. It is hard to think about empirical analysis in macroeconomics today without comprehensive systems of national accounts. General equilibrium theory, as formulated by Arrow and Debreu, has created a general *theoretical* system helping us to grasp the idea of the interaction of billions of economic transactions in millions of different markets. Without the modern system of national accounts, however, we could not obtain an *empirical registration* of these transactions in comprehensive aggregates. The idea of national accounts harks back over several centuries, and theoretical and empirical work on national accounts flourished in the 1930s, as reflected in the works by Ragnar Frisch, Erik Lindahl, Colin Clark and Simon Kuznets. But Richard Stone was the leading architect of the modern *system* of national accounts, which married the principles of macroeconomic bookkeeping and aggregate macroeconomic models. Leontief-style input-output tables also became a useful component of this type of work.

These methodological prizes referred to advances in empirical analysis. Methodological contributions in theory have also been rewarded. One example is the shared Prize to Leonid Kantorovich and Tjalling Koopmans (in 1975). Kantorovich defined, as early as 1939, the concept of efficient resource use in individual enterprises and later developed similar efficiency conditions for the economy as a whole. He also demonstrated the theoretical connection between the allocation of resources and the price system, both at a certain point of time and in a growing economy. Koopmans' so-called activity analysis, in a similar vein, clarified the correspondence between efficiency in production and existence of a system of 'accounting prices'. Both showed how the theoretical possibility of decentralized decision-making in a planned economy is connected with the existence of an efficient price system, including a uniform accounting price of capital on which to base investment decisions. This analysis was, in fact, closely related to the earlier discussed achievements in general equilibrium theory by Arrow and Debreu. Though both laureates have also made important contributions to the mathematical technique of linear programming, this was not what they were honored for; instead they received the Prize for enriching our understanding of basic economic issues in normative allocative theory by applying new tools of analysis.

One of the most important theoretical methods development in recent decades is game theory. While John von Neumann and Oskar Morgenstern did pioneering work in this field as early as the late 1940s, the analytical breakthrough was spawned by John Harsanyi, John Nash and Reinhard Selten, who were awarded (in 1994) "for their pioneering analysis of equilibrium in the theory of non-cooperative games."

John Nash introduced the distinction between cooperative games, in which binding agreements can be made, and non-cooperative games, where binding agreements are not feasible. Nash also developed an equilibrium concept for predicting the outcome of non-cooperative games that later came to be called the Nash equilibrium. Reinhard Selten was the first to refine the Nash equilibrium concept for analyzing dynamic strategic interaction among different agents and to apply these refinements in the analyses of competition with only a few sellers. These refinements made it possible to exclude a number of theoretically possible but unstable or irrelevant equilibria. John Harsanyi showed how games can be rigorously analyzed in the case of incomplete information. In this way he provided a theoretical foundation for predicting the outcome of strategic interaction between agents imperfectly informed, for instance, about the objectives of other individuals. Hence, Harsanyi gave an important impetus to further development in the field of information economics, after the pioneering work of Stigler, Vickrey and Mirrlees.

2. Problems and Difficulties

What then are the main problems and difficulties in choosing Laureates in economics? It may be useful to discuss this issue in connection with four questions: (a) How should 'economics' be interpreted in the context of the awards? (b) What criteria should be used when judging whether a candidate merits a prize? (c) In what order should worthy candidates be selected? (d) When and for what reason should prizes be shared?

2.1. *The Scope of Economics*

The prize committee, and the Academy, has decided to give wide interpretation to the term 'economic sciences', so that prizes may be awarded to scholars making important scientific contributions also in neighboring disciplines, in so far as these concern economic issues. In other words, 'interdisciplinary research' has been regarded as important. Indeed, as mentioned above, several awards have been given for contributions on the borderline between economics, political science, sociology and history.

Scholars with traditional training in economics have increasingly been 'trespassing' into neighboring territory by applying the methods of economic theory and econometrics to problems not previously analyzed much by economists. These various trespassing tendencies have led George Stigler [1], as well as other economists, to talk about economics as 'The Imperial Science'. It is also true, however, that research in other social sciences has recently influenced research in economics.

Though the Academy, and its selection committee, has followed the same general principles as applied to the prizes in the natural sciences, i.e. to award specific contributions, the degree of 'specificity' of the awards has varied considerably. Examples of prizes with high specificity are the awards to Wassily Leontief, Ragnar Frisch, Jan Tinbergen, Trygve Haavelmo, James Heckman, Daniel McFadden, James Mirrlees and William Vickrey, as well as the prizes to game theory and financial economics. The smallest degree of specificity is probably found in the prize citations for Paul Samuelson, Milton Friedman, Friedrich von Hayek, Gunnar Myrdal and Amartya Sen. In the case of Paul Samuelson reference was made to his contribution to "raising the level of analysis in economic science." The prize citation to Milton Friedman mentioned his contributions to consumption analysis and to monetary history and theory as well as "his demonstration of the complexity of stabilization policy." The latter referred to Friedman's stress on how time lags, conflicts of goals, uncertainty and endogenous expectations among economic agents greatly complicates stabilization policy. In the prize citation for Gunnar Myrdal and Friedrich von Hayek, the Academy mentioned both their "pioneering work in the theory of money and economic fluctuations" and "their penetrating analysis of the interdependence of economic, social and institutional phenomena." In the case of Amartya Sen, much of the contributions referred to his clarification of the philosophical foundations of economics. His empirical studies of starvation in a number of poor countries integrated political and sociological factors with more narrowly economic ones. Simon Kuznets was awarded for his lifetime contributions to the empirical analyses of economic development. Thus, the Academy has awarded not only narrowly defined specific contributions but also clusters of such contributions, including lifetime achievements if these consist of major contributions to economic science widely interpreted.

2.2. *Criteria for Awards*

When considering what should be regarded as a 'worthy' contribution, it is probably correct to say that the selection committee has looked, in particular, at the *originality* of the contribution, its scientific and practical *importance*, and its *impact* on scientific work. To provide shoulders on which other scholars can stand, and thus climb higher, has been regarded as an important contribution. To some extent, the committee has also considered the impact on society at large, including the impact on public policy.

An issue is whether the contributions by a scholar should be treated as gross or net. In other words, should the prize-awarding authority make deductions for 'bad' (low-quality) research? It is obvious that no such deductions have been made. Moreover, how does one deal with people who, in addition to their scholarly work, have participated in the political debate with policy recommendations which sometimes may reflect strong ideological commitments? Friedman, Hayek, Myrdal, Tinbergen, Tobin, Modigliani, Solow and Mundell are obvious examples. In conformity with the basic idea of the Prize as a scientific award, such activities have been neglected.

When deciding who should be regarded as worthy of a prize, the scrutiny of time has helped the committee considerably. Because the Prize was initiated as late as 1969, time has sorted out worthy candidates, for whom the risk of 'premature fame' is minimal. During the first decade of the Economics Prize, the committee largely had the task of working with a heavy backlog of rather obvious candidates. Indeed, some of the honored contributions were made several decades ago, even as far back as the thirties, examples being the awards to Frisch, Tinbergen, Hicks, Ohlin and Kantorovich.

Moreover, it usually takes a longer time in economics (and social sciences in general) than in the natural sciences to find out if a new contribution is *solid* or if it is just a fad. In other words, it is important to wait for scrutiny, criticism and repeated tests of the quality and relevance of a contribution. The reason is not only that economic behavior, like human behavior in general, is complex but also that it varies over time and place. This is partly because individuals learn from previous experience, which may make empirically estimated behavior patterns unstable. Thus, new results may turn out to be relevant only to a transient conjuncture of circumstances, having much less generality than was supposed at first. Another reason to be particularly careful is that relevant empirical tests usually take time to pursue, partly because such tests usually rely on non-experimental data.

When trying to define a prizeworthy contribution, the selection committee has not relied much on quantitative indicators such as the number of nominations or the frequency of citations, even though the prize-winners usually rank very high on both accounts[a] [2, 3]. Indeed, there are a number of exceptions of prize-winners who have received quite a few nominations and who also rank quite low in citation indices, pronounced examples being Kantorovich, Stone, Haavelmo, Allais, Meade and Ohlin (though the Heckscher–Ohlin model is often referred to in the literature). There are also some economists who consistently rank very high on citation indices, but who have not received prizes.

2.3. *The Order of Awards*

What, then, have been the main criteria for choosing the order of worthy candidates? There is an unavoidable subjectivity and arbitrariness in this choice. Two dominant criteria seem to have been: (i) to give early prizes to particularly important contributions, and (ii) to adhere to a pluralist view of economic research, by shifting over the years between candidates in different fields, using different methods of analysis, and reflecting different views of the world. There has also been (iii) a tendency to give prizes in chronological order of discovery.

2.4. *Sharing of Prizes*

Another important issue is when, and how, awards should be shared. According to the rules laid down for the Nobel Prizes, the Prize can be shared among a maximum of three persons. A shared Nobel Prize is just as honorable as a single prize, and each Laureate has to be worthy of the Prize alone.

For receiving a shared award, there has to be some 'common denominator' among the Laureates. Shared awards in economics have been given *either* when the contributions are the results of actual collaborative work, *or* when the contributors are so closely related that a sharing is important to demonstrate the connection and to be 'fair' to contributors. So far, eleven

[a]For instance, in a study by Richard Quandt (1976), based on citations in eight leading journals, Nobel Memorial Prize-winners constitute 13 out of 21 highest ranked (then living) economists in terms of citations in 1960 (with the six top economists all, by now, having been prize-winners). They constitute 11 out of the 26 highest ranked in citations in 1970 (with the top five all having been awarded by now).

prizes out of thirty have been shared, which is somewhat less frequent than in the natural sciences during the last three decades.

The prize-awarding authority has interpreted the common denominator of shared prizes in economics in different ways for different awards. For instance, the contributions of Ragnar Frisch and Jan Tinbergen were strongly linked by intellectual influence, in particular from the older Laureate (Frisch) to the younger. The shared Prize between John R. Hicks and Kenneth Arrow also reflected the work of two different generations working in the same field, more specifically in general equilibrium and welfare theory. In the words of the press release of the Academy, Hicks 'initiated' a profound transformation of general equilibrium theory, while Arrow 'provided it with fresh nourishment'. The Prize in game theory was also an award to two generations of contributions, with Nash being a pioneer and Harsanyi and Selten making Nash's concepts of non-cooperative game theory more applicable.

The Prize shared between Tjalling Koopmans and Leonid Kantorovich reflected instead similarity of mutually *independent* contributions in the field of normative economic theory, or more specifically a normative theory of the optimum allocation of resources. The shared prizes in the theory of information economics to Vickrey and Mirrlees were of a similar nature.

The prize–sharing between Hayek and Myrdal was, again, of a different nature. Both were pioneers in macro and monetary analysis in the thirties — the Austrian School and the Stockholm School, respectively. They both used the concepts of aggregate savings and investment to explain macroeconomic fluctuations. Both later broadened the scope of economic analysis, by emphasizing the institutional, legal, political and ideological framework of economic and social processes. The fact that they are often regarded as political 'antipoles' did not bother the committee, since the Prize is a purely scientific award. This is probably the shared award for which the common denominator of the Laureates' achievements was the smallest.

Some shared prizes have instead been awards for complementary contributions. The common denominator for the shared Prize to Bertil Ohlin and James Meade was their analysis of international trade and capital movements. The contributions of Arthur Lewis and Theodore Schultz were also largely complementary. The common denominator is that their research has dealt with long-term economic development for less developed countries. Another complementary price was the shared award between Fogel and North, which was designed to honor the two most important pioneers in 'new' economic history, in which modern tools of economic and statistical analysis are applied to issues in economic history.

The shared Prize to Markowitz, Miller and Sharpe was also an award for complementary contributions, in this case in financial economics, though the latter two had the advantage of standing on the shoulders of Markowitz. The Prize to Merton and Scholes may be regarded as a 'follow up' of this Prize, since they (with the late Fisher Black) developed a theory of price formation for one specific type of important financial asset, namely 'derivative financial instruments', such as options and futures. This is one of the clearest cases of a 'joint' contribution in the sense that the Laureates cooperated in the research that led to their achievements. The prize in microeconometrics to Heckman and McFadden was also shared for complementary contributions.

3. Do the Prizes Reflect New Trends in Economic Analysis?

The awards that have been made so far obviously reflect some characteristic features of economic analysis during the last half-century. First of all, the awards clearly reflect the dominant role of the United States in economic research during this period. Out of 46 laureates, 30 have been United States citizens. However, although all of these had been working in the United States for a long time, it may be worth noting that four of them — Leontief, Koopmans, Debreu and Harsanyi — were born and largely trained in other countries. The only other countries that have received prizes (as defined by citizenship) are the United Kingdom (6 awards), Sweden and Norway (2 awards each), France, Canada, India, the Netherlands and the Soviet Union (one each). The only universities where faculty members have received more than a single award are Chicago (9 awards), Harvard (4 awards), Cambridge (4 awards), MIT (3 awards), Berkeley (3 awards), and Columbia, Oslo, Princeton, Stanford and Yale (2 awards each).[b]

Turning to the content of the awarded contributions, the emphasis on deductive rather than inductive methods in economic analysis shows up strongly. The increased role of mathematical formulation is also strongly reflected in the awards, important examples being the prizes to Samuelson, Hicks, Arrow, Koopmans, Kantorovich, Debreu, Allais, as well as the laureates in financial economics and game theory.

[b]Of course, a Laureate may have made his main contribution before he joined the institution where he was affiliated when awarded the Prize.

Another characteristic trend in economics during the second half of the 20th century is the growing importance of quantitative methods including systematic statistical testing or estimation, i.e. econometrics. This development is reflected notably in the awards to Frisch, Tinbergen, Leontief, Klein, Stone, Haavelmo, Heckman and McFadden. Indeed, the huge volume of quantitative research during the last decade, often involving large masses of data, would hardly have been possible without the development of analytical techniques such as econometrics, input-output analysis, programming, as well as the development of powerful computers.

The awards also illustrate the important role of macroeconomics during the postwar period (in particular, Friedman, Klein, Tobin, Modigliani, Solow, Lucas and Mundell). New ways of looking at the economic system have also been recognized by the awarding authority, as reflected in the awards to economics of information, human capital and game theory as well as the role of economic institutions.

A final but difficult question: Has the selection committee viewed the award as a chance to influence the direction of new research in economics? The answer is 'no' in the sense that the committee has tried to be broad and pluralistic of outlook in its decisions about awards, and to emphasize the multidimensional nature of economic research. Somewhat paradoxically, such an eclectic approach could, of course, be regarded in itself as a way to influence views about fruitful research, by recognizing research fields and methods that may not for the moment be the focus of interest.

It may also be argued that the prize-awarding authority has demonstrated that there are many different ways to advance a science like economics: rigorous deductive theorizing, whether by way of verbal or mathematical techniques; the development and application of new concepts and methods of analysis; rigorous empirical testing of existing hypotheses, as well as less formalized confrontation of various hypotheses with empirical fact; or 'simply' profound observation and nonformalized innovative thinking about economic issues.

Bibliography

Stigler, G., *Economics — The Imperial Science?*, Mimeo, April 1984.

Quandt, R., "Some Quantitative Aspects of the Economic Journal Literature," *Journal of Political Economy*, Aug. 1976, 84 (4, Part 1), 741–755.

Grubel, H. G., "Citation Counts for Leading Economists," *Economic Notes by Monte Dei Paschi Di Siena*, 1979, 2, 134–145.

Ståhle, N. K., *Alfred Nobel and the Nobel Prizes*, Stockholm: The Nobel Foundation and The Swedish Institute, 1978.

Economic Sciences 1970

Paul A. Samuelson (1915–)

"for the scientific work through which he has developed static and dynamic economic theory and actively contributed to raising the level of analysis in economic science"

Economic Sciences 1988

Maurice Allais (1911–)

"for his pioneering contributions to the theory of markets and efficient utilization of resources"

Economic Sciences 1998

Amartya Sen (1933–)

"for his contributions to welfare economics"

LIST OF NOBEL LAUREATES (1901–2000)

Year	Physics	Chemistry	Physiology or Medicine
1901	W.C. Röntgen (G)	J.H. van't Hoff (NL)	E.A. von Behring (G)
1902	H.A. Lorentz (NL) P. Zeeman (NL)	H.E. Fischer (G)	R. Ross (GB)
1903	A.H. Becquerel (F) P. Curie (F) M. Curie (F)	S.A. Arrhenius (Swe)	N.R. Finsen (D)
1904	J.W.S. Rayleigh (GB)	W. Ramsey (GB)	I.P. Pavlov (R)
1905	P.E.A. von Lenard (G)	J.F.W.A. von Baeyer (G)	R. Koch (G)
1906	J.J. Thomson (GB)	H. Moissan (F)	C. Golgi (I) S. Ramón y Cajal (Sp)
1907	A.A. Michelson (US)	E. Buchner (G)	C.L.A. Laveran (F)
1908	G. Lippman (F)	E. Rutherford (GB)	I.I. Mechnikov (R) P. Ehrlich (G)
1909	G. Marconi (I) C.F. Braun (G)	W. Ostwald (G)	E.T. Kocher (Swi)
1910	J. D. van der Waals (NL)	O. Wallach (G)	A. Kossel (G)
1911	W. Wien (G)	M. Curie (F)	A. Gullstrand (Swe)
1912	N.G. Dalén (Swe)	V. Grignard (F) P. Sabatier (F)	A. Carrel (F)
1913	H. Kamerlingh-Onnes (NL)	A. Werner (Swi)	C.R. Richet (F)
1914	M. von Laue (G)	T.W. Richards (US)	R. Bárány (Au)
1915	W.H. Bragg (GB) L.W. Bragg (GB)	R.M. Willstätter (G)	Not awarded
1916	Not awarded	Not awarded	Not awarded
1917	C.G. Barkla (GB)	Not awarded	Not awarded

Year	Literature	Peace
1901	Sully Prudhomme (F)	J.H. Dunant (Swi) F. Passy (F)
1902	Theodor Mommsen (G)	E. Ducommun (Swi) C.A. Gobat (Swi)
1903	Bjørnstjerne Bjørnson (N)	W.R. Cremer (GB)
1904	Frédéric Mistral (F) José Echegaray (Sp)	Institute of International Law, Ghent
1905	Henryk Sienkiewicz (Pol)	B.S.F. von Suttner (Au)
1906	Giosué Carducci (I)	T. Roosevelt (US)
1907	Rudyard Kipling (GB)	E.T. Moneta (I) L. Renault (F)
1908	Rudolf Eucken (G)	K.P. Arnoldson (Swe) F. Bajer (D)
1909	Selma Lagerlöf (Swe)	A.M.F. Beernaert (B) P.H.B.B. d'Estournelles (F)
1910	Paul Heyse (G)	Permanent International Peace Bureau, Berne
1911	Maurice Maeterlinck (B)	T.M.C. Asser (NL) A.H. Fried (Au)
1912	Gerhart Hauptmann (G)	E. Root (US)
1913	Rabindranath Tagore (In)	H. La Fontaine (B)
1914	Not awarded	Not awarded
1915	Romain Rolland (F)	Not awarded
1916	Verner von Heidenstam (Swe)	Not awarded
1917	Karl Gjellerup (D) Henrik Pontoppidan (D)	International Committee of the Red Cross, Geneva

Year	Physics	Chemistry	Physiology or Medicine
1918	M.K.E.L. Planck (G)	F. Haber (G)	Not awarded
1919	J. Stark (G)	Not awarded	J. Bordet (B)
1920	C.E. Guillaume (Swi)	W.H. Nernst (G)	S.A.S. Krogh (D)
1921	A. Einstein (G/Swi)	F. Soddy (GB)	Not awarded
1922	N.H.D. Bohr (D)	F.W. Aston (GB)	A.V. Hill (GB) O.F. Meyerhof (G)
1923	R.A. Millikan (US)	F. Pregl (Au)	F.G. Banting (Ca) J.J.R. Macleod (Ca)
1924	K.M.G. Siegbahn (Swe)	Not awarded	W. Einthoven (NL)
1925	J. Franck (G) G.L. Hertz (G)	R.A. Zsigmondy (G)	Not awarded
1926	J.B. Perrin (F)	T. Svedberg (Swe)	J.A.G. Fibiger (D)
1927	A.H. Compton (US) C.T.R. Wilson (GB)	H.O. Wieland (G)	J. Wagner-Jauregg (Au)
1928	O.W. Richardson (GB)	A.O.R. Windaus (G)	C.J.H. Nicolle (F)
1929	L-V.P.R. de Broglie (F)	A. Harden (GB) H.K.A.S. von Euler-Chelpin (Swe)	C. Eijkman (NL) F.G. Hopkins (GB)
1930	C.V. Raman (In)	H. Fischer (G)	K. Landsteiner (Au)
1931	Not awarded	C. Bosch (G) F. Bergius (G)	O.H. Warburg (G)
1932	W.K. Heisenberg (G)	I. Langmuir (US)	C.S. Sherrington (GB) E.D. Adrian (GB)
1933	E. Schrödinger (Au) P.A.M. Dirac (GB)	Not awarded	T.H. Morgan (US)
1934	Not awarded	H.C. Urey (US)	G.H. Whipple (US) G.R. Minot (US) W.P. Murphy (US)
1935	J. Chadwick (GB)	F. Joliot (F) I. Joliot-Curie (F)	H. Spemann (G)
1936	V.F. Hess (Au) C.D. Anderson (US)	P.J.W. Debye (NL)	H.H. Dale (GB) O. Loewi (Au)

Year	Literature	Peace
1918	Not awarded	Not awarded
1919	Carl Spitteler (Swi)	T.W. Wilson (US)
1920	Knut Hamsun (N)	L.V.A. Bourgeois (F)
1921	Anatole France (F)	K.H. Branting (Swe) C.L. Lange (N)
1922	Jacinto Benavente (Sp)	F. Nansen (N)
1923	W.B. Yeats (Ir)	Not awarded
1924	Wladyslaw Reymont (Pol)	Not awarded
1925	G.B. Shaw (GB)	A. Chamberlain (GB) C.G. Dawes (US)
1926	Grazia Deledda (I)	A. Briand (F) G. Stresemann (G)
1927	Henri Bergson (F)	F. Buisson (F) L. Quidde (G)
1928	Sigrid Undset (N)	Not awarded
1929	Thomas Mann (G)	F.B. Kellogg (US)
1930	Sinclair Lewis (US)	L.O.N. Söderblom (Swe)
1931	Erik Axel Karlfeldt (Swe)	J. Addams (US) N.M. Butler (US)
1932	John Galsworthy (GB)	Not awarded
1933	Ivan Bunin (stateless)	N. Angell (GB)
1934	Luigi Pirandello (I)	A. Henderson (GB)
1935	Not awarded	C. von Ossietzky (G)
1936	Eugene O'Neill (US)	C. Saavedra Lamas (Ar)

Year	Physics	Chemistry	Physiology or Medicine
1937	C.J. Davisson (US) G.P. Thomson (GB)	W.N. Haworth (GB) P. Karrer (Swi)	A. von Szent-Györgyi Nagyrapolt (H)
1938	E. Fermi (I)	R. Kuhn (G)	C.J.F. Heymans (B)
1939	E.O. Lawrence (US)	A.F.J. Butenandt (G) L. Ružička (Swi)	G. Domagk (G)
1940	Not awarded	Not awarded	Not awarded
1941	Not awarded	Not awarded	Not awarded
1942	Not awarded	Not awarded	Not awarded
1943	O. Stern (US)	G. de Hevesy (H)	H.C.P. Dam (D) E.A. Doisy (US)
1944	I.I. Rabi (US)	O. Hahn (G)	J. Erlanger (US) H.S. Gasser (US)
1945	W. Pauli (Au)	A.I. Virtanen (Fi)	A. Fleming (GB) E.B. Chain (GB) H.W. Florey (GB)
1946	P.W. Bridgman (US)	J.B. Sumner (US) J.H. Northrop (US) W.M. Stanley (US)	H.J. Muller (US)
1947	E.V. Appleton (GB)	R. Robinson (GB)	C.F. Cori (US) G.T. Cori (US) B.A. Houssay (Ar)
1948	P.M.S. Blackett (GB)	A.W.K. Tiselius (Swe)	P.H. Müller (Swi)
1949	H. Yukawa (J)	W.F. Giauque (US)	W.R. Hess (Swi) A.C. de Abreu Freire Egas Moniz (Por)
1950	C.F. Powell (GB)	O.P.H. Diels (FRG) K. Alder (FRG)	E.C. Kendall (US) T. Reichstein (Swi) P.S. Hench (US)
1951	J.D. Cockcroft (GB) E.T.S. Walton (Ir)	E.M. McMillan (US) G.T. Seaborg (US)	M. Theiler (SA)
1952	F. Bloch (US) E.M. Purcell (US)	A.J.P. Martin (GB) R.L.M. Synge (GB)	S.A. Waksman (US)
1953	F. Zernike (NL)	H. Staudinger (FRG)	H.A. Krebs (GB) F.A. Lipmann (US)

Year	Literature	Peace
1937	Roger Martin du Gard (F)	Viscount Cecil of Chelwood (GB)
1938	Pearl Buck (US)	Nansen International Office for Refugees, Geneva
1939	F.E. Sillanpää (Fi)	Not awarded
1940	Not awarded	Not awarded
1941	Not awarded	Not awarded
1942	Not awarded	Not awarded
1943	Not awarded	Not awarded
1944	Johannes V. Jensen (D)	International Committee of the Red Cross, Geneva
1945	Gabriela Mistral (Chile)	C. Hull (US)
1946	Hermann Hesse (Swi)	E.G. Balch (US) J.R. Mott (US)
1947	André Gide (F)	The Friends Service Council (GB) The American Friends Service Committee (US)
1948	T.S. Eliot (GB)	Not awarded
1949	William Faulkner (US)	Lord Boyd Off of Brechin (GB)
1950	Bertrand Russell (GB)	R. Bunche (US)
1951	Pär Lagerkvist (Swe)	L. Jouhaux (F)
1952	François Mauriac (F)	A. Schweitzer (F/G)
1953	Winston Churchill (GB)	G.C. Marshall (US)

Year	Physics	Chemistry	Physiology or Medicine
1954	M. Born (GB) W. Bothe (FRG)	L. C. Pauling (US)	J.F. Enders (US) T.H. Weller (US) F.C. Robbins (US)
1955	W.E. Lamb (US) P. Kusch (US)	V. du Vigneaud (US)	A.H.T. Theorell (Swe)
1956	W.B. Shockley (US) J. Bardeen (US) W.H. Brattain (US)	C.N. Hinshelwood (GB) N.N. Semenov (USSR)	A.F. Cournand (US) W. Forssmann (FRG) D.W. Richards, Jr. (US)
1957	C.N. Yang (China) T-D. Lee (China)	A.R. Todd (GB)	D. Bovet (I)
1958	P.A. Cherenkov (USSR) I.M. Frank (USSR) I.Y. Tamm (USSR)	F. Sanger (GB)	G.W. Beadle (US) E.L. Tatum (US) J. Lederberg (US)
1959	E.G. Segrè (US) O. Chamberlain (US)	J. Heyrovsky (Cz)	S. Ochoa (US) A. Kornberg (US)
1960	D.A. Glaser (US)	W.F. Libby (US)	F.M. Burnet (Aus) P.B. Medawar (GB)
1961	R. Hofstadter (US) R.L. Mössbauer (FRG)	M. Calvin (US)	G. von Békésy (US)
1962	L.D. Landau (USSR)	M.F. Perutz (GB) J.C. Kendrew (GB)	F.H.C. Crick (GB) J.D. Watson (US) M.H.F. Wilkins (GB)
1963	E.P. Wigner (US) M. Goeppert-Mayer (US) J.H.D. Jensen (FRG)	K. Ziegler (FRG) G. Natta (I)	J.C. Eccles (Aus) A.L. Hodgkin (GB) A.F. Huxley (GB)
1964	C.H. Townes (US) N.G. Basov (USSR) A.M. Prokhorov (USSR)	D. Crowfoot Hodgkin (GB)	K. Bloch (US) F. Lynen (FRG)
1965	S-I. Tomonaga (J) J. Schwinger (US) R.P. Feynman (US)	R.B. Woodward (US)	F. Jacob (F) A. Lwoff (F) J. Monod (F)
1966	A. Kastler (F)	R.S. Mulliken (US)	P. Rous (US) C.B. Huggins (US)
1967	H.A. Bethe (US)	M. Eigen (FRG) R.G.W. Norrish (GB) G. Porter (GB)	R. Granit (Swe) H.K. Hartline (US) G. Wald (US)

Year	Literature	Peace
1954	Ernest Hemingway (US)	Office of the UN High Commissioner for Refugees, Geneva
1955	Halldór Laxness (Ic)	Not awarded
1956	J.R. Jiménez (Sp)	Not awarded
1957	Albert Camus (F)	L.B. Pearson (Ca)
1958	Boris Pasternak (USSR) (declined the prize)	G. Pire (B)
1959	Salvatore Quasimodo (I)	P.J. Noel-Baker (GB)
1960	Saint-John Perse (F)	A.J. Luthuli (SA)
1961	Ivo Andrić (Y)	D.H.A.C. Hammarskjöld (Swe)
1962	John Steinbeck (US)	L.C. Pauling (US)
1963	Giorgos Seferis (Gr)	International Committee of the Red Cross, Geneva League of Red Cross Societies, Geneva
1964	Jean-Paul Sartre (F) (declined the prize)	M.L. King, Jr. (US)
1965	Michail Sholokhov (USSR)	United Nations Children's Fund (UNICEF)
1966	Shmuel Y. Agnon (Is) Nelly Sachs (FRG)	Not awarded
1967	Miguel A. Asturias (Guat)	Not awarded

Year	Physics	Chemistry	Physiology or Medicine
1968	L.W. Alvarez (US)	L. Onsager (US)	R.W. Holley (US) H.G. Khorana (US) M.W. Nirenberg (US)
1969	M. Gell-Mann (US)	D.H.R. Barton (GB) O. Hassel (N)	M. Delbrück (US) A.D. Hershey (US) S.E. Luria (US)
1970	H.O.G. Alfvén (Swe) L.E.F. Néel (F)	L.F. Leloir (Ar)	B. Katz (GB) U. von Euler (Swe) J. Axelrod (US)
1971	D. Gabor (GB)	G. Herzberg (Ca)	E.W. Sutherland, Jr. (US)
1972	J. Bardeen (US) L.N. Cooper (US) J.R. Schrieffer (US)	C.B. Anfinsen (US) S. Moore (US) W.H. Stein (US)	G.M. Edelman (US) R.R. Porter (GB)
1973	L. Esaki (J) I. Giaever (US) B.D. Josephson (GB)	E.O. Fischer (FRG) G. Wilkinson (GB)	K. von Frisch (FRG) K. Lorenz (Au) N. Tinbergen (GB)
1974	M. Ryle (GB) A. Hewish (GB)	P.J. Flory (US)	A. Claude (B) C. de Duve (B) G.E. Palade (US)
1975	A.N. Bohr (D) B.R. Mottelson (D) L.J. Rainwater (US)	J.W. Cornforth (Aus/GB) V. Prelog (Swi)	D. Baltimore (US) R. Dulbecco (US) H.M. Temin (US)
1976	B. Richter (US) S.C.C. Ting (US)	W.N. Lipscomb (US)	B.S. Blumberg (US) D.C. Gajdusek (US)
1977	P.W. Anderson (US) N.F. Mott (GB) J.H. Van Vleck (US)	I. Prigogine (B)	R. Guillemin (US) A.V. Schally (US) R. Yalow (US)
1978	P.L. Kapitsa (USSR) A.A. Penzias (US) R.W. Wilson (US)	P. Mitchell (GB)	W. Arber (Swi) D. Nathans (US) H.O. Smith (US)
1979	S.L. Glashow (US) A. Salam (Pak) S. Weinberg (US)	H.C. Brown (US) G. Wittig (FRG)	A.M. Cormack (US) G.N. Hounsfield (GB)
1980	J.W. Cronin (US) V.L. Fitch (US)	P. Berg (US) W. Gilbert (US) F. Sanger (GB)	B. Benacerraf (US) J. Dausset (F) G.D. Snell (US)

Year	Literature	Peace	Economic Sciences
1968	Yasunari Kawabata (J)	R. Cassin (F)	
1969	Samuel Beckett (Ir)	International Labour Organization, Geneva	R. Frisch (N) J. Tinbergen (NL)
1970	Aleksandr Solzhenitsyn (USSR)	N.E. Borlaug (US)	P.A. Samuelson (US)
1971	Pablo Neruda (Chile)	W. Brandt (FRG)	S. Kuznets (US)
1972	Heinrich Böll (FRG)	Not awarded	J.R. Hicks (GB) K.J. Arrow (US)
1973	Patrick White (Aus)	H.A. Kissinger (US) Le Duc Tho (V) (declined the prize)	W. Leontief (US)
1974	Eyvind Johnson (Swe) Harry Martinson (Swe)	S. Mac Bride (Ir) E. Sato (J)	G. Myrdal (Swe) F.A. von Hayek (GB)
1975	Eugenio Montale (I)	A.D. Sakharov (USSR)	L.V. Kantorovich (USSR) T.C. Koopmans (US)
1976	Saul Bellow (US)	B. Williams (GB) M. Corrigan (GB)	M. Friedman (US)
1977	Vicente Aleixandre (Sp)	Amnesty International, London	B. Ohlin (Swe) J.E. Meade (GB)
1978	Isaac B. Singer (US)	M.A. al-Sadat (Eg) M. Begin (Is)	H.A. Simon (US)
1979	Odysseus Elytis (Gr)	Mother Teresa (In)	T.W. Schultz (US) A. Lewis (GB)
1980	Czesław Miłosz (Pol/US)	A. Pérez Esquivel (Ar)	L.R. Klein (US)

Year	Physics	Chemistry	Physiology or Medicine
1981	N. Bloembergen (US) A.L. Schawlow (US) K.M. Siegbahn (Swe)	K. Fukui (J) R. Hoffmann (US)	R.W. Sperry (US) D.H. Hubel (US) T.N. Wiesel (Swe)
1982	K.G. Wilson (US)	A. Klug (GB)	S. Bergström (Swe) B. Samuelsson (Swe) J.R. Vane (GB)
1983	S. Chandrasekhar (US) W.A. Fowler (US)	H. Taube (US)	B. McClintock (US)
1984	C. Rubbia (I) S. van der Meer (NL)	R.B. Merrifield (US)	N.K. Jerne (D) G.J.F. Köhler (FRG) C. Milstein (GB/Ar)
1985	K. von Klitzing (FRG)	H.A. Hauptman (US) J. Karle (US)	M.S. Brown (US) J.L. Goldstein (US)
1986	E. Ruska (FRG) G. Binnig (FRG) H. Rohrer (Swi)	D.R. Herschbach (US) Y.T. Lee (US) J.C. Polanyi (Ca)	S. Cohen (US) R. Levi-Montalcini (I/US)
1987	J.G. Bednorz (FRG) K.A. Müller (Swi)	D.J. Cram (US) J-M. Lehn (F) C.J. Pedersen (US)	S. Tonegawa (J)
1988	L.M. Lederman (US) M. Schwartz (US) J. Steinberger (US)	J. Deisenhofer (FRG) R. Huber (FRG) H. Michel (FRG)	J. W. Black (GB) G.B. Elion (US) G.H. Hitchings (US)
1989	N.F. Ramsey (US) H.G. Dehmelt (US) W. Paul (FRG)	S. Altman (US/Ca) T.R. Cech (US)	J.M. Bishop (US) H.E. Varmus (US)
1990	J.I. Friedman (US) H.W. Kendall (US) R.E. Taylor (Ca)	E.J. Corey (US)	J.E. Murray (US) E.D. Thomas (US)
1991	P-G. de Gennes (F)	R.R. Ernst (Swi)	E. Neher (G) B. Sakmann (G)
1992	G. Charpak (F)	R.A. Marcus (US)	E.H. Fischer (US/Swi) E.G. Krebs (US)
1993	R.A. Hulse (US) J.H. Taylor, Jr. (US)	K.B. Mullis (US) M. Smith (Ca)	R.J. Roberts (GB) P.A. Sharp (US)

Year	Literature	Peace	Economic Sciences
1981	Elias Canetti (GB)	Office of the UN High Commissioner for Refugees, Geneva	J. Tobin (US)
1982	Gabriel García Márquez (Co)	A. Myrdal (Swe) A. García Robles (M)	G.J. Stigler (US)
1983	William Golding (GB)	L. Wałesa (Pol)	G. Debreu (US)
1984	Jaroslav Seifert (Cz)	D.M. Tutu (SA)	R. Stone (GB)
1985	Claude Simon (F)	International Physcians for the Prevention of Nuclear War, Boston	F. Modigliani (US)
1986	Wole Soyinka (Ni)	E. Wiesel (US)	J.M. Buchanan, Jr. (US)
1987	Joseph Brodsky (US)	O. Arias Sanchez (CR)	R.M. Solow (US)
1988	Naguib Mahfouz (Eg)	The United Nations Peace-Keeping Forces	M. Allais (F)
1989	Camilo José Cela (Sp)	The 14th Dalai Lama (Tenzin Gyatso) (T)	T. Haavelmo (N)
1990	Octavio, Paz (M)	M. Gorbachev (USSR)	H.M. Markowitz (US) M.H. Miller (US) W.F. Sharpe (US)
1991	Nadine Gordimer (SA)	Aung San Suu Kyi (Burma)	R.H. Coase (GB)
1992	Derek Walcott (St. Lucia)	R. Menchú Tum (Guat)	G.S. Becker (US)
1993	Toni Morrison (US)	N. Mandela (SA) F.W. de Klerk (SA)	R.W. Fogel (US) D.C. North (US)

Year	Physics	Chemistry	Physiology or Medicine
1994	B.N. Brockhouse (Ca) C.G. Shull (US)	G.A. Olah (US)	A.G. Gilman (US) M. Rodbell (US)
1995	M.L. Perl (US) F. Reines (US)	P.J. Crutzen (NL) M.J. Molina (US) F.S. Rowland (US)	E.B. Lewis (US) C. Nüsslein-Volhard (G) E.F. Wieschaus (US)
1996	D.M. Lee (US) D.D. Osheroff (US) R.C. Richardson (US)	R.F. Curl, Jr. (US) H.W. Kroto (GB) R.E. Smalley (US)	P.C. Doherty (Aus) R.M. Zinkernagel (Swi)
1997	S. Chu (US) C. Cohen-Tannoudji (F) W.D. Phillips (US)	P.D. Boyer (US) J.E. Walker (GB) J.C. Skou (D)	S.B. Prusiner (US)
1998	R.B. Laughlin (US) H.L. Störmer (G) D.C. Tsui (US)	W. Kohn (US) J.A. Pople (GB)	R.F. Furchgott (US) L.J. Ignarro (US) F. Murad (US)
1999	G. 't Hooft (NL) M.J.G. Veltman (NL)	A.H. Zewail (Eg/US)	G. Blobel (US)
2000	Z.I. Alferov (R) H. Kroemer (G) J.S. Kilby (US)	A.J. Heeger (US) A.G. MacDiarmid (US) H. Shirakawa (J)	A. Carlsson (Swe) P. Greengard (US) E.R. Kandel (US)

Year	Literature	Peace	Economic Sciences
1994	Kenzaburo Oe (J)	Y. Arafat (Pal) S. Peres (Is) Y. Rabin (Is)	J.C. Harsanyi (US) J.F. Nash (US) R. Selten (G)
1995	Seamus Heaney (Ir)	J. Rotblat (GB) Pugwash Conferences on Science and World Affairs (Ca)	R.E. Lucas, Jr. (US)
1996	Wisława Szymborska (Pol)	C.F. Ximenes Belo (East Timor) J. Ramos-Horta (East Timor)	J.A. Mirrlees (GB) W. Vickrey (US)
1997	Dario Fo (I)	International Campaign to Ban Landmines Jody Williams (US)	R.C. Merton (US) M.S. Scholes (US)
1998	José Saramago (Por)	J. Hume (N Ir) D. Trimble (N Ir)	A. Sen (In)
1999	Günter Grass (G)	Médecins Sans Frontières	R.A. Mundell (Ca)
2000	Gao Xingjian (F)	Kim Dae Jung (South Korea)	J.J. Heckman (US) D.L. McFadden (US)

Abbreviations

Ar Argentina; **Aus** Australia; **Au** Austria; **B** Belgium; **Ca** Canada; **Co** Colombia; **CR** Costa Rica; **Cz** Czechoslovakia; **D** Denmark; **Eg** Egypt; **Fi** Finland; **F** France; **FRG** Federal Republic of Germany; **G** Germany (before 1948 and after 1990); **GB** Great Britain; **Gr** Greece; **Guat** Guatemala; **H** Hungary; **Ic** Iceland; **In** India; **Ir** Ireland; **Is** Israel; **I** Italy; **J** Japan; **M** Mexico; **NL** The Netherlands; **Ni** Nigeria; **N Ir** Northern Ireland; **N** Norway; **Pak** Pakistan; **Pal** Palestine; **Pol** Poland; **Por** Portugal; **R** Russia (between 1922 and 1991 USSR); **Sp** Spain; **Swe** Sweden; **Swi** Switzerland; **SA** South Africa; **T** Tibet; **US** United States; **V** Vietnam; **Y** Yugoslavia.

DISCLAIMER

CARMEL CLAY PUBLIC LIBRARY
3 1690 00945 5008